Brain Informatics and Health

Editors-in-Chief

Ning Zhong, Department of Life Science and Informatics, Maebashi Institute of Technology, Maebashi-City, Japan

Ron Kikinis, Department of Radiology, Harvard Medical School, Boston, USA

Series Editors

Weidong Cai, School of Computer Science, The University of Sydney, Sydney, Australia

Henning Müller, University of Applied Sciences Western Switzerland, Sierre, Switzerland

Hirotaka Onoe, Graduate School of Medicine, Kyoto University, Kobe, Japan

Sonia Pujol, Department of Radiology, Harvard Medical School, Boston, USA

Philip S. Yu, Department of Computer Science, University of Illinois at Chicago, Chicago, USA

Informatics-enabled studies are transforming brain science. New methodologies enhance human interpretive powers when dealing with big data sets increasingly derived from advanced neuro-imaging technologies, including fMRI, PET, MEG, EEG and fNIRS, as well as from other sources like eye-tracking and from wearable, portable, micro and nano devices. New experimental methods, such as in to imaging, deep tissue imaging, opto-genetics and dense-electrode recording are generating massive amounts of brain data at very fine spatial and temporal resolutions. These technologies allow measuring, modeling, managing and mining of multiple forms of big brain data. Brain informatics & health related techniques for analyzing all the data will help achieve a better understanding of human thought, memory, learning, decision-making, emotion, consciousness and social behaviors. These methods also assist in building brain-inspired, human-level wisdom-computing paradigms and technologies, improving the treatment efficacy of mental health and brain disorders.

The Brain Informatics & Health (BIH) book series addresses the computational, cognitive, physiological, biological, physical, ecological and social perspectives of brain informatics as well as topics relating to brain health, mental health and well-being. It also welcomes emerging information technologies, including but not limited to Internet of Things (IoT), cloud computing, big data analytics and interactive knowledge discovery related to brain research. The BIH book series also encourages submissions that explore how advanced computing technologies are applied to and make a difference in various large-scale brain studies and their applications.

The series serves as a central source of reference for brain informatics and computational brain studies. The series aims to publish thorough and cohesive overviews on specific topics in brain informatics and health, as well as works that are larger in scope than survey articles and that will contain more detailed background information. The series also provides a single point of coverage of advanced and timely topics and a forum for topics that may not have reached a level of maturity to warrant a comprehensive textbook.

Wanzeng Kong · Xuanyu Jin

Brain Fingerprint Identification

 Springer

Wanzeng Kong
School of Computer Science
Hangzhou Dianzi University
Hangzhou, Zhejiang, China

Xuanyu Jin
School of Computer Science
Hangzhou Dianzi University
Hangzhou, Zhejiang, China

ISSN 2367-1742 ISSN 2367-1750 (electronic)
Brain Informatics and Health
ISBN 978-981-96-4511-4 ISBN 978-981-96-4512-1 (eBook)
https://doi.org/10.1007/978-981-96-4512-1

This book was supported by the National Natural Science Foundation of China (62471169, U20B2074), Key Research and Development Project of Zhejiang Province (2023C03026), and the Key Laboratory of Brain Machine Collaborative Intelligence of Zhejiang Province (2020E10010).

© The Editor(s) (if applicable) and The Author(s) 2025. This book is an open access publication.

Open Access This book is licensed under the terms of the Creative Commons Attribution-NonCommercial-NoDerivatives 4.0 International License (http://creativecommons.org/licenses/by-nc-nd/4.0/), which permits any noncommercial use, sharing, distribution and reproduction in any medium or format, as long as you give appropriate credit to the original author(s) and the source, provide a link to the Creative Commons license and indicate if you modified the licensed material. You do not have permission under this license to share adapted material derived from this book or parts of it.

The images or other third party material in this book are included in the book's Creative Commons license, unless indicated otherwise in a credit line to the material. If material is not included in the book's Creative Commons license and your intended use is not permitted by statutory regulation or exceeds the permitted use, you will need to obtain permission directly from the copyright holder.

This work is subject to copyright. All commercial rights are reserved by the author(s), whether the whole or part of the material is concerned, specifically the rights of translation, reprinting, reuse of illustrations, recitation, broadcasting, reproduction on microfilms or in any other physical way, and transmission or information storage and retrieval, electronic adaptation, computer software, or by similar or dissimilar methodology now known or hereafter developed. Regarding these commercial rights a non-exclusive license has been granted to the publisher.

The use of general descriptive names, registered names, trademarks, service marks, etc. in this publication does not imply, even in the absence of a specific statement, that such names are exempt from the relevant protective laws and regulations and therefore free for general use.

The publisher, the authors and the editors are safe to assume that the advice and information in this book are believed to be true and accurate at the date of publication. Neither the publisher nor the authors or the editors give a warranty, expressed or implied, with respect to the material contained herein or for any errors or omissions that may have been made. The publisher remains neutral with regard to jurisdictional claims in published maps and institutional affiliations.

This Springer imprint is published by the registered company Springer Nature Singapore Pte Ltd.
The registered company address is: 152 Beach Road, #21-01/04 Gateway East, Singapore 189721, Singapore

If disposing of this product, please recycle the paper.

Foreword by Andrzej Cichocki

In 2018, during a conference at the RIKEN Institute in Japan, I had the pleasure of attending Prof. Kong's presentation on *Brain Fingerprint Identification*. His talk on using EEG signals for biometric authentication immediately piqued my interest. The concept of non-stealable and tamper-resistant identification based on brain patterns seemed like a revolutionary leap in security.

What intrigued me most during that presentation was the unique advantages that brain signals offer for secure identification. The inherent complexity of EEG data, combined with the sophisticated analytical methods Prof. Kong has developed, opened up exciting new possibilities in areas requiring high levels of security and precision. His ability to address the challenges of working with time-varying, noisy EEG data was especially impressive, and it became clear that his research would have far-reaching implications across multiple fields.

Since that initial encounter, I have closely followed Prof. Kong's research, which has continued to push the boundaries of brain signal analysis and its practical applications. His work provides deep insights into the potential of EEG-based identification systems, and this book reflects the significant progress that has been made in recent years. It is an important contribution to the field and will undoubtedly be an essential resource for anyone looking to understand and engage with this promising area of research.

I highly recommend the book to researchers, practitioners, and students interested in the convergence of neuroscience, machine learning, and biometric security. Professor Kong's work in this field is groundbreaking. This book is an invaluable resource for anyone seeking to explore the future of brain-based personal identification.

January 2025

Andrzej Cichocki
Polish Academy of Sciences
Warsaw, Poland

Andrzej Cichocki (Fellow, IEEE) received his M.Sc. (with honors), Ph.D., and Dr.Sc. (Habilitation) degrees in electrical engineering from Warsaw University of Technology, Warsaw, Poland, in 1972, 1976, and 1982, respectively. He spent several years at the University of Erlangen-Nuremberg, Germany, as an Alexander von Humboldt Research Fellow and guest professor. He was also the senior team leader and head of the Laboratory for Advanced Brain Signal Processing at the RIKEN Brain Science Institute, Japan. From 2017 to 2021, he received a prestigious research mega-grant from Skoltech. He is currently a professor at the Systems Research Institute, Polish Academy of Sciences.

Foreword by Fabio Babiloni

I am pleased to write this Foreword for Prof. Kong's remarkable book on *Brain Fingerprint Identification*. Having followed the development of this innovative technology for some time, I am convinced that its potential to transform the field of biometric security is vast.

One of the most compelling aspects of Brain Fingerprint Identification is its ability to provide continuous authentication. Unlike traditional biometric systems that authenticate a person at a single point in time, Brain Fingerprint Identification takes advantage of the continuous nature of EEG signals. This not only enhances the security of the authentication process but also opens up exciting possibilities for applications in real-world settings, where constant verification is essential. The uninterrupted flow of brain data enables ongoing, real-time identity verification, making it an ideal solution for high-security environments that require persistent monitoring.

Professor Kong's works in this field have made significant strides toward overcoming the challenges inherent in EEG-based identification, such as session variability and cognitive task dependence. This book explores these issues in depth while proposing novel solutions that ensure the robustness and reliability of brain fingerprint systems.

Brain fingerprint identification will play an increasingly important role in the future, not only for security applications but also in areas such as healthcare, personalized technology, and even everyday user interaction with devices. This book offers a detailed yet accessible guide to understanding the science behind brainprints and their practical applications. I highly recommend it to researchers, professionals, and students eager to explore the future of continuous, secure personal identification.

January 2025

Fabio Babiloni
Sapienza University of Rome
Rome, Italy

Fabio Babiloni (Member of the Academy of Europe) is a professor of Physiology at the Faculty of Medicine of the University of Rome "La Sapienza," Rome, Italy. He is also a professor of Biomedical Engineering. He is the author of more than 250 papers on bioengineering and neurophysiological topics in international peer-reviewed scientific journals, and more than 250 contributions to conferences and book chapters. He is the associate editor of IEEE TBME and IEEE TNSRE and editor-in-chief of the International Journal of Bioelectromagnetism. His total impact factor is more than 500, and his H-index is 63. He is an associate editor of IEEE Transactions on Biomedical Engineering, IEEE Transactions on Neural System and Rehabilitation Engineering, and it has been in the AdCom of IEEE-EMBS since 2012.

Preface

Brain fingerprint identification is a novel biometric technology based on EEG signals, offering advantages such as non-stealability, continuous identification, anti-spoofing detection, and resistance to coercion. These features make it particularly suitable for high-security domains, such as the military and finance. Thanks to machine learning techniques, which are effective at capturing both the features of the data and the latent dependencies between those features, significant progress has been made in brain fingerprint identification.

However, EEG signals are characterized by low signal-to-noise ratios and time-varying properties, which result in distribution differences in EEG data collected at different time points, leading to instability in brainprint features extracted across different sessions. Additionally, existing brain fingerprint identification methods often require subjects to perform specific cognitive tasks, which limits the practical applicability in real-world scenarios. Finally, EEG signals are influenced by a variety of factors, and the nervous system exhibits continuous spontaneous variability. This makes it difficult for brain fingerprint identification models to perform well in scenarios involving unknown sessions or cognitive tasks, presenting a challenge for the generalization of such models.

This book addresses these challenges by exploring brain fingerprint identification in three key research areas: specific-task multi-session identification, multi-task specific-session identification, and multi-task multi-session identification. Chapter 1 introduces the background and related works on brain fingerprint identification. Chapter 2 provides a detailed explanation of the generation and acquisition of EEG signals. Chapters 3 and 4 discuss the study of specific-task multi-session brain fingerprint identification, using novel graph neural networks and domain-adversarial learning-based disentangled representation methods. Chapters 5–8 focus on multi-task single-session EEG brain fingerprint identification, utilizing machine learning methods such as brain networks and low-rank sparse decomposition, as well as deep learning techniques like residual networks, multi-scale convolutional neural networks, and tensor-train neural networks to extract more robust, task-independent brainprint features. Chapter 9 explores multi-task multi-session brain fingerprint identification using an attention network based on domain adaptation learning.

Chapter 10 presents research on cross-task and cross-session brain fingerprint identification, innovatively employing the disentangled adversarial generalization network to extract identity-related features from EEG signals that are independent of mental tasks and data acquisition differences. Chapters 11 and 12 provide a summary of the book and offer a forward-looking perspective on the future of brain fingerprint identification, respectively. In this book, "multi-session" and "multi-task" refer to the involvement of different sessions or mental tasks in the model training process, while "cross-session" and "cross-task" refer to scenarios where the EEG data used for validation or testing corresponds to session or tasks not included in the training phase.

Brain fingerprint identification based on EEG signals is an emerging approach in the field of biometric identification. Our goal is to introduce this promising research area to interested readers. The methods presented in this book have demonstrated optimal performance in experiments and provide new research ideas and technical support for the development of stable and generalizable brain fingerprint identification applications. This book can serve as a reference for researchers, practitioners, and professionals, and can also be used as an introductory textbook for senior undergraduate students.

Since brain fingerprint identification based on EEG signals is still in its early stages, we hope that this book will offer an overview of the field and lay a foundation for further exploration in both theoretical and practical aspects.

Due to time and knowledge limitations, we acknowledge that there may be numerous issues and errors. We kindly ask readers to point out these mistakes for correction. We sincerely hope you enjoy reading this book and join us in exploring this promising and uncharted field with us.

Hangzhou, China Wanzeng Kong
January 2025 Xuanyu Jin

Acknowledgements This book is the culmination of our extensive research into EEG-based brain fingerprint identification. We extend our deepest gratitude to our colleagues and friends for their invaluable assistance, advice, insights, and guidance throughout this journey. Special thanks go to the members and collaborating researchers at the Key Laboratory of Brain Machine Collaborative Intelligence of Zhejiang Province. Their contributions provided a solid foundation for the content presented in this book. We would like to acknowledge the significant contributions made to several chapters by our team members: Honggang Liu (Chaps. 3, 4), Bei Jiang and Qiaonan Fan (Chap. 5), Xianghao Kong (Chap. 6), Yuxuan Zhu (Chap. 7), Xinyu Yang (Chap. 9), and Ni Li (Chap. 10). Additionally, our appreciation extends to Menghang Li, Xinran Wang, and Jing Tang for their meticulous proofreading and enhancements to the manuscript. We are also immensely grateful to the professional team at Springer, especially Dr. Nick Zhu, for their patience and unwavering support throughout the publication process.

This book was supported by the National Natural Science Foundation of China (62471169, U20B2074), Key Research and Development Project of Zhejiang Province (2023C03026), and the Key Laboratory of Brain Machine Collaborative Intelligence of Zhejiang Province (2020E10010).

Competing Interests The authors have no competing interests to declare that are relevant to the content of this manuscript.

Contents

1	**Overall of Brain Fingerprint Identification**		1
	1.1 Background		1
	1.2 Related Work		4
		1.2.1 Brain Fingerprint Identification for Specific-Task and Specific-Session EEG	4
		1.2.2 Brain Fingerprint Identification for Multi-session EEG	6
		1.2.3 Brain Fingerprint Identification for Multi-task EEG	8
		1.2.4 Brain Fingerprint Identification for Multi-session and Multi-task EEG	9
		1.2.5 Summary of Related Work	10
	References		12
2	**Basics of EEG Signals**		15
	2.1 Generation of EEG Signals		15
		2.1.1 EEG Signals Based on Exogenous Stimuli	16
		2.1.2 EEG Signals Based on Endogenous Stimulation	18
	2.2 Acquisition of EEG Signals		20
	2.3 EEG Dataset for Brain Fingerprint Identification		22
		2.3.1 Specific-Task and Specific-Session EEG Dataset	22
		2.3.2 Specific-Task and Multi-session EEG Dataset	24
		2.3.3 Multi-task and Specific-Session EEG Dataset	27
		2.3.4 Multi-task and Multi-session EEG Dataset	28
	References		30
3	**Specific-Task and Multi-session Brain Fingerprint Identification with Multi-scale Graph Neural Network**		33
	3.1 Introduction		33
	3.2 Multi-scale Convolution and Graph Pooling Network		35
		3.2.1 Dynamic Feature Extraction Module	35
		3.2.2 Graph-Level Pooling Module	37

		3.2.3	Graph-Level Attention Embedding Module	38
		3.2.4	Optimization	38
	3.3	Brain Fingerprint Identification with MCGP		40
		3.3.1	Data Pre-processing	40
		3.3.2	Comparison with Existing Methods	41
		3.3.3	Implementation Details	41
		3.3.4	Performances for Mixed Affective State Scenario	42
		3.3.5	Performance in the Single Affective State Scenario	43
		3.3.6	Effect of Each Component	44
	3.4	Conclusion		47
	References			47
4	**Specific-Task and Multi-session Brain Fingerprint Identification with Joint Disentangled Representation**			**49**
	4.1	Introduction		49
	4.2	Joint Disentangled Representation with Domain Adversarial Training Framework		51
		4.2.1	Feature Extractor	52
		4.2.2	Representation Disentanglement	53
	4.3	Brain Fingerprint Identification with JDR-DAT		55
		4.3.1	Data Pre-processing	55
		4.3.2	Implementation Details	56
		4.3.3	Performances of Specific-Task and Multi-session Brain Fingerprint Identification	56
		4.3.4	Effect of Each Component	59
		4.3.5	Effect of Critical Frequency Bands	60
	4.4	Conclusion		60
	References			61
5	**Multi-task and Single-Session Brain Fingerprint Identification with Brain Network**			**63**
	5.1	Introduction		64
	5.2	Brain Network		65
		5.2.1	Pre-processing	65
		5.2.2	Phase Synchronization	66
		5.2.3	Construction of Brain Network	68
		5.2.4	Linear Discriminant Analysis	70
	5.3	Brain Fingerprint Identification with Brain Network		72
		5.3.1	Attributes Analysis	72
		5.3.2	Performances of Multi-task and Single-Session Brain Fingerprint Identification	77
		5.3.3	Effect of Four EEG Bands	78
		5.3.4	Effect of Combined Attributes	79
	5.4	Conclusion		80
	References			81

Contents

6 Multi-task and Single-Session Brain Fingerprint Identification with Low-Rank and Matrix Decomposition 83
- 6.1 Introduction 83
- 6.2 Low-Rank and Matrix Decomposition 85
  - 6.2.1 Domain Transformation of EEG Signals 86
  - 6.2.2 LRMD Model-Based Background EEG Extraction 88
  - 6.2.3 Ensemble Subspace Construction 91
  - 6.2.4 Reconstruction Coefficient Generation 91
  - 6.2.5 MCC-Based Classification 92
- 6.3 Brain Fingerprint Identification with LRMD 92
  - 6.3.1 Implementation Details 93
  - 6.3.2 Performances of Multi-task and Single-Session Brain Fingerprint Identification 93
  - 6.3.3 Performances of LRMD with Different Kernels 96
- 6.4 Conclusion 97
- References 98

7 Multi-task and Single-Session Recognition with Residual Multi-scale Neural Network 101
- 7.1 Introduction 101
- 7.2 Residual and Multi-scale Spatio-Temporal Convolution Neural Network 103
  - 7.2.1 Residual Learning 103
  - 7.2.2 Multi-scale Grouping Convolution 104
  - 7.2.3 Global Average Pooling 105
  - 7.2.4 Batch Normalization 106
  - 7.2.5 Structure of Network 107
  - 7.2.6 Training Settings 108
- 7.3 Brain Fingerprint Identification with RAMST-CNN 108
  - 7.3.1 Description of Datasets 108
  - 7.3.2 Data Pre-processing 109
  - 7.3.3 Performances of Multi-task and Single-Session Brain Fingerprint Identification 110
  - 7.3.4 Effect of Each Component 113
- 7.4 Conclusion 114
- References 114

8 Multi-task and Single-Session with Convolutional Tensor-Train Neural Network 117
- 8.1 Introduction 118
- 8.2 Convolutional Tensor-Train Neural Network 119
  - 8.2.1 Tensor Network 119
  - 8.2.2 Convolutional Tensor-Train Neural Network Architecture 119
- 8.3 Brain Fingerprint Identification with CTNN 122
  - 8.3.1 EEG Datasets 122

		8.3.2	Data Pre-processing	122
		8.3.3	Experiment Setup	123
		8.3.4	Comparison of the Performance on Different Models	123
		8.3.5	Comparison of Multi-task Brain Fingerprint Identification with Different TT-Ranks	126
		8.3.6	Comparison of Multi-task Brain Fingerprint Identification Among EEG from Different Frequency Bands	127
		8.3.7	Comparison of Multi-task Brain Fingerprint Identification with Different Channels Selections	127
	8.4	Conclusion		130
	References			134
9	Multi-task and Multi-session Brain Fingerprint Identification with Attention Neural Network with Domain Adaptation Learning			135
	9.1	Introduction		136
	9.2	Tensorized Spatial-Frequency Attention Network		137
		9.2.1	Preliminaries	138
		9.2.2	Intra-Source Transferable Feature Learning	139
		9.2.3	Tensorized Spatial-Frequency Attention Mechanism	141
	9.3	Brain Fingerprint Identification with TSFAN		143
		9.3.1	Data Pre-processing	143
		9.3.2	Baseline Methods	144
		9.3.3	Implementation Details	144
		9.3.4	Performance Comparison of TSFAN with Baseline Methods	145
		9.3.5	Effect of Tensorized Spatial-Frequency Attention Mechanism	151
		9.3.6	Effect of Tensor Ranks for Tucker Network	151
		9.3.7	Effect of Trade-Off Parameter λ and γ	153
		9.3.8	Effect of Electrodes for EEG-Based Biometric Recognition	154
	9.4	Conclusion		157
	References			158
10	Cross-Task and Cross-Session Brain Fingerprint Identification with Disentangled Adversarial Generalization Network			161
	10.1	Introduction		162
	10.2	Disentangled Adversarial Generalization Network		163
		10.2.1	Backbone Convolutional Neural Network	165
		10.2.2	Feature Disentanglement	166

		10.2.3	Attention Mechanism with Adversarial Self-Challenging	167
	10.3	Brain Fingerprint Identification with DAGN		169
		10.3.1	Data Description and Pre-processing	170
		10.3.2	Baseline Approaches	171
		10.3.3	Implementation Details	172
		10.3.4	Comparison of Performance with State-of-the-Art Models	172
		10.3.5	Ablation Experiments for the Disentangled Adversarial Generalization Network	175
		10.3.6	Effect of Sample Length	178
		10.3.7	Effect of the Different Frequency Bands	180
		10.3.8	Effect of the Number of Electrodes	182
	10.4	Conclusion		184
	References			185
11	**Summary**			187
12	**Future Directions**			189

Chapter 1
Overall of Brain Fingerprint Identification

Abstract This chapter provides a comprehensive overview of brain fingerprint identification, a cutting-edge technique that utilizes electroencephalography (EEG) to analyze brain activity patterns for personal identification. It begins with an in-depth background on the evolution of biometric systems, emphasizing the limitations of traditional methods such as fingerprint, facial recognition, and iris scans. The chapter highlights the unique advantages of brain fingerprint identification, including its potential for high accuracy and security, as well as its ability to operate in real time without physical contact. Following the background, the chapter delves into related work, reviewing significant studies and advancements in the field of brain fingerprint identification. This section examines various methodologies employed in existing research and discusses how these methods contribute to the effectiveness of personal identification. It highlights the unique advantages of using brain activity patterns, captured through electroencephalography (EEG), for identity verification in comparison to traditional biometric methods. Finally, the chapter concludes with a critical summary of the related work, identifying existing gaps in the literature and the challenges that need to be addressed for brain fingerprint identification to achieve widespread adoption. By highlighting these challenges, this chapter sets the groundwork for the subsequent research presented in the book, emphasizing the need for further exploration into the technological advancements, and potential applications of brain fingerprint identification.

1.1 Background

With the rapid development of digitization and informatization, many individuals and organizations opt to store their vast amounts of data on network or cloud platforms. However, due to the increasingly complex network environment, the threats to network security are becoming more challenging, necessitating higher security measures for effectively managing data information. In response to this demand, the concept of Zero Trust Security, based on the principle of "never trust, always verify", has gradually garnered attention from researchers. Within this framework, secure identity management and real-time dynamic access control serve as the cornerstone for building the entire Zero Trust Security system. Each entity needs to obtain corresponding access permissions through identity management. Simultaneously, all individuals and events are regarded as untrusted, requiring continuous monitoring and identity authentication of all entities to adjust the real-time access permissions for different entities, thereby enhancing the security of information management.

Identity management typically involves identification and verification. Early identity management often employed usernames and passwords, providing a certain degree of authentication for information systems. However, these methods carry significant risks of theft and forgetfulness. According to the General Data Protection Regulation of the European Union, biometric identification based on individuals' unique physiological or behavioral characteristics is a highly secure and effective identity management. Compared to early identity management technologies, biometric features are more complex and do not require individual memory, effectively reducing the risks of identity theft and unauthorized access, thus better protecting personal privacy. Consequently, current identity management technologies increasingly adopt biometric identification techniques to strengthen identity verification and enhance system security. The most widely used biometric features include face, fingerprint, palmprint, and iris, as well as behavioral characteristics such as handwritten signatures, gait, and keystrokes.

However, although existing biometrics are already highly regarded, significant risks are still associated with them. Physiological features such as fingerprints, irises, and faces are susceptible to theft, replication, and tampering. Current technologies can extract individual fingerprint and facial information from images and utilize 3D printing to create corresponding replicas for identity information decryption. Furthermore, with the advancement of deep learning technology, synthetically generated realistic facial images have been used to deceive facial identification systems [1, 2]. Additionally, behavioral characteristics such as signatures and gait are prone to imitation and attacks in public surveillance environments. Considering the limitations of traditional biometric identification technologies, researchers at Binghamton University have proposed a novel biometric feature named brainprint [3] for brain fingerprint identification based on electroencephalogram (EEG) signals. This technology aims to analyze signals generated by brain neuron activities to extract individual identity information. Compared to existing traditional biometric features, brainprint features based on EEG signals offer the following advantages: **Non-stealability**: Unlike external details such as facial features, fingerprints, and gait, which are susceptible to exposure in monitoring environments leading to information leakage, EEG signals are intrinsic physiological characteristics with high concealment. Moreover, from a physiological perspective, EEG signals reflect the manifestation of brain activities, enabling them to reveal individual unique neural pathway patterns, making it difficult for attackers to forge or replicate through physiological mimicry. **Continuous identification**: Individual neural activity patterns possess robust identity discriminability over time [4–6]. Additionally, due to the continuity of EEG signals, brain print features can be employed to continuously verify user identity without being limited to a single authentication. This is more suitable for scenarios that require uninterrupted

1.1 Background

dynamic identity management. **Anti-spoofing detection**: Research has shown that only the brains of individuals with vital signs can generate EEG signals. In addition, different stimulus-evoked modes will lead to changes in the activity patterns of different brain regions, thus generating corresponding EEG signals. Therefore, the system's resistance to attack can be enhanced by decoding the EEG signals under stimulus-evoked changes for in anti-spoofing detection along with continuous authentication **Resistance to coercion**: Individuals are prone to anxiety, fear, and other emotions under coercion, leading to differences in the EEG signals of individuals in different states. Therefore, detecting the abnormal state of individual EEG signals makes it possible to reject coercive user authentication while realizing effective identity recognition.

With these characteristics, the brainprint feature is suitable for applications with high-security requirements such as military, finance, and information security, thus attracting increasing attention [7, 8]. In the metaverse and virtual reality spaces, brainprint is emerging as a novel tool for user authentication. With the increasing reliance on virtual environments for communication, commerce, and entertainment, securing user identity and access to personal data is paramount. brainprint can enable seamless and secure authentication, where users' brain activity serves as the key for logging into their virtual profiles or engaging in secure transactions. Similarly, in healthcare, brain fingerprint identification has potential applications in patient identification, especially in environments where high levels of privacy are required, such as mental health clinics or hospitals with sensitive data. It can ensure that access to patient records, medication, and treatment history is restricted to the rightful individual, enhancing both privacy and safety. Moreover, brainprint can be integrated into financial systems to verify user identities during transactions, potentially reducing fraud and identity theft. In forensic investigations, the application of brainprint could assist in identifying individuals based on cognitive responses to specific stimuli, helping to solve criminal cases where traditional identification methods fall short. As brain fingerprint identification technology continues to evolve, its applications will expand, providing a more secure, reliable, and personalized way of verifying identity across a range of sectors.

In existing research on brain fingerprint identification methods, traditional machine learning approaches require specialized prior knowledge for feature extraction, making it difficult to cope with the complexity and temporal variability of EEG signals, resulting in limited model identification accuracy and generalization ability [9]. In contrast to machine learning, deep neural networks can effectively capture the high-level features of the data and the potential dependencies between the features, and effectively learn the feature representations in complex EEG signals. Therefore, significant progress has been made in brain fingerprint identification based on EEG signals. Various types of deep learning methods, such as Convolutional Neural Networks (CNN), Recurrent Neural Networks (RNN), Graph Convolutional Neural Networks (GCNN), etc., have been shown to extract identity-related brainprint features from EEG signals.

1.2 Related Work

Due to the low signal-to-noise ratio and temporal variability of EEG signals, the extraction of brainprint features is susceptible to the influence of different EEG acquisition sessions and individual mental states. Therefore, existing brain fingerprint identification research can be categorized into four categories based on various sessions and mental tasks: brain fingerprint identification for EEG signals from specific sessions and tasks, brain fingerprint identification for EEG signals from multiple sessions EEG, brain fingerprint identification for EEG signals from different tasks EEG, and brain fingerprint identification for EEG signals from numerous sessions and tasks EEG. This section will provide detailed introductions to the four categories of existing brain fingerprint identification research mentioned above.

1.2.1 Brain Fingerprint Identification for Specific-Task and Specific-Session EEG

The acquisition of EEG signals is usually based on designed experimental paradigms that specify the cognitive tasks and processes individuals need to perform during EEG acquisition. EEG collection paradigms based on specific mental tasks can activate specific brain areas and corresponding neural activity patterns in individuals to analyze characters associated with specific mental tasks. According to different EEG acquisition paradigms, existing brain fingerprinting is based on the following mental states: resting state, exogenous stimulation task, and endogenous spontaneous task, respectively.

In brain fingerprint identification based on resting state, individuals need to maintain wakefulness and relaxation during EEG signal acquisition, without external stimuli or execution of any endogenous spontaneous mental tasks. This paradigm typically includes two resting states: eyes closed and eyes open. Resting-state EEG provides the most essential and fundamental information about brain activity. It is often used as a baseline task for EEG signal acquisition, making it relatively easy to obtain. A single-session identity recognition study involving 109 participants showed that in both eyes-closed and eyes-open resting states, the accuracy of identity recognition reached 100%, with average error rates ranging from 0.044 to 0.0196% [10–12]. Moctezuma et al. [13] utilized EEG data from only three electrodes and employed Discrete Wavelet Transformation (DWT) and Empirical Mode Decomposition (EMD) to decompose EEG signals and extract identity-related features, achieving the highest correct acceptance rates of 99.3 and 99.7% for the two resting states, respectively. Ma et al. [14] designed a convolutional neural network model consisting of two convolutional layers, two pooling layers, and a fully connected layer. The identification accuracies of 10 participants under eyes-open and eyes-closed conditions were 88 and 86%, respectively.

1.2 Related Work

In brain fingerprint identification based on exogenous stimulus tasks, Event-Related Potentials (ERP) are a special type of external stimulus-induced task, characterized by positive potential waves generated by the brain cortex in response to certain stimuli perceived by individuals. Koike-Akino et al. [15] employed various machine learning dimensionality reduction methods on EEG data from 25 participants in a P300 evoked potential task, achieving an identification accuracy of 96.7%. Based on P300 potential, Rathi et al. [16] proposed a novel visual stimulus-based identity verification method using images as selection attributes. They employed classification methods such as Quadratic Discriminant Analysis, k-nearest Neighbors, and Quadratic Support Vector Machine, achieving an identification accuracy of up to 97% for 10 participants. Steady-state Visual Evoked Potential (SSVEP) is another paradigm based on external visual stimuli, where the brain cortex produces electric potentials at the same frequency as the flickering frequency of the visual stimuli when participants gaze at fixed-frequency visual stimuli. Due to the high signal-to-noise ratio of SSVEP signals, they have been used for identity recognition. Zhang et al. [17] investigated individual coherence and phase synchronization under different visual stimuli by analyzing frequency components in different frequency bands. They employed the Support Vector Machine, Random Forest, and k-nearest Neighbors algorithms for 15 participants, achieving a high identification accuracy of up to 98%. Additionally, Rapid Serial Visual Presentation (RSVP) is another visual stimulus-based paradigm where participants are presented with a series of consecutive visual stimuli, such as words, images, or symbols, rapidly presented on a screen. Participants must pay attention to specific targets in this continuous sequence of presentations. Currently, the RSVP paradigm is also used for identity recognition. Chen et al. [18] recorded and analyzed EEG signals from 29 participants under the RSVP paradigm using wet and dry electrodes, achieving a true acceptance rate of 100% for identity recognition with both wet and dry electrodes, with average login times of 13.5 and 27 s, respectively. Zhang et al. [19] utilized a self and non-self-name stimulus paradigm for EEG data collection, achieving good Brainprint identification performance through a combination of multi-scale convolutional structures and various authentication strategies.

Unlike brain fingerprint identification which relies on exogenous tasks, brain fingerprint identification based on endogenous spontaneous tasks requires individuals to generate mental activities spontaneously, without specific external stimuli. This typically involves intrinsic thought processes driven by the individual's internal mental activities, such as imagination and reasoning. For motor imagery tasks, Nguyen et al. [20] employed speech feature extraction methods, such as Mel-Frequency Cepstral Coefficients, to process EEG signals, achieving identity recognition rates of 99 and 80.8% for 3 and 9 participants, respectively. Wu et al. [21] analyzed the causal relationships of signals using Partial Directed Coherence, achieving an average error rate of <1% for the identification among 109 participants through convolutional neural networks. Bidgoly et al. [22] reduced the number of EEG electrode channels to three using the Gram-Schmidt orthogonalization process while maintaining an accuracy rate of 99.9%. Moctezuma et al. [23] conducted identity recognition for 20 participants imagining Spanish words, achieving an identification accuracy of 92% by combining EMD and Linear Support Vector Machine.

Moreover, Wilaiprasitporn et al. [24] researched brain fingerprint identification based on affective EEG. They designed a model that combines convolutional neural networks and recurrent neural networks for identity recognition, achieving a maximum average correct identity recognition rate of 100% among 32 participants. Li et al. [25] proposed a novel tensor method for brain fingerprint identification, which includes tensor representation construction, tensor learning methods such as Generalized Tensor Discriminant Analysis and tensor metric learning, as well as a tensor measurement method based on Manhattan distance. They verified the proposed method on three affective EEG datasets, achieving identification accuracy of over 99%.

In conclusion, despite the remarkable progress made in brain fingerprint identification based on specific sessions and tasks, challenges remain. Existing methods require individuals to demonstrate high levels of cooperation and concentration, lacking generalization across different mental tasks. Furthermore, brain fingerprint identification methods based on deep neural networks necessitate numerous samples for model training, which are often difficult to obtain in real-world scenarios, leading to unreliable models.

1.2.2 Brain Fingerprint Identification for Multi-session EEG

In practical application scenarios, users' enrollment and verification data typically come from different collection times. This requires brainprint recognition methods to maintain high performance and accuracy for EEG data collected across different sessions. Therefore, it is essential to validate the stability of individual identity-related components in brain neural activity patterns across different sessions. Da Silva et al. [4] utilized magnetoencephalography (MEG) technology, while Van De Ville et al. [5] employed functional magnetic resonance imaging (fMRI) to verify that individual-specific neural activity patterns exhibit robust identity discriminability across different collection sessions. Shenghao et al. [6] found that the accuracy of individual identity recognition based on time and frequency features far exceeds random chance levels. Additionally, they observed consistent individual identity features across different times and tasks using both MEG and EEG data, demonstrating consistency across various neural imaging modalities.

In brain fingerprint identification based on resting-state EEG, La Rocca et al. [26] employed an autoregressive statistical model of reflectance coefficients and tested it with a linear classifier. They achieved identity recognition accuracy of up to 100% for nine subjects during specific periods of eyes-open and eyes-closed conditions. In contrast, the cross-session identity recognition accuracies were 75.72 and 97.9%, respectively, using the same electrode channel setup. Ma et al. [27] proposed a method combining non-stationary multivariate empirical mode decomposition with intrinsic mode correlation features of brain connectivity and conducted identity recognition

1.2 Related Work

using machine learning classifiers. The F1 performance of brain fingerprint identification for 81 subjects during specific sessions reached 99%, while for 32 subjects across sessions, the F1 performance dropped to only 55%. Nakamura et al. [28] utilized features such as power spectral density and autoregressive model coefficients. In mixed validation across two-time periods for 15 subjects, the half-total error rate was 6.9% with an identification accuracy of 98.3%. For cross-session validation, the half-total error rate was 17.2%, and the accuracy was 95.7%. Maiorana [29] demonstrated that by combining convolutional neural networks and recurrent neural networks, the identity recognition accuracies for 40 subjects during eyes-closed and eyes-open resting states were 91.4 and 81.9%, respectively.

In brain fingerprint identification based on external stimulus tasks, the N400 event-related potential triggered by single-word reading, associated with semantic memory, has been utilized for identity recognition. Armstrong et al. [3] employed various pattern classifiers, including support vector machines, cross-correlation, divergence autoencoders, and naive discriminant learning. The results showed that the cross-session identification accuracy averaged 89% over an average interval of 12.73 d, while the average identification accuracy over intervals as long as six months reached 93%. For visually evoked potential paradigms, Piciucco et al. [30] proposed an identification system based on SSVEP, using Mel-frequency cepstral coefficients and autoregressive reflection coefficients as brainprint features. In cross-session validation with an average interval of 15 d, the average correct identification rate for 25 subjects was 91.47%. Zhang et al. [31] discovered that individuals exhibit specific frequency sensitivity to SSVEP visual stimuli in the 7–12 Hz range and further explored the permanency of the proposed SSVEP features. Through multiple experiments conducted at five different time points (morning and afternoon of the same day, one-day interval, one-week interval, and one-month interval) with 15 subjects, the results indicated that the proposed individual brainprint features were replicable. Zeng et al. [32] introduced an RSVP identity authentication paradigm based on facial images, integrating facial and EEG features to extract more specific and stable features for identity verification. They proposed an authentication method based on hierarchical discriminant component analysis and a genetic algorithm to optimize model construction for each subject. This method achieved an average authentication accuracy of 94.26% within 6 s. Even with an average time interval of 30 d, an average accuracy of 88.88% was maintained. Zhao et al. [33] designed a novel encoding-modulated visual-evoked potential paradigm for brain fingerprint identification, employing canonical correlation analysis and task-related component analysis to identify target individuals. With a single time window of 5.25 s of EEG data, they achieved a 100% correct identification rate, and the cross-session identification rate for two sessions was 99.43%. Furthermore, auditory evoked potentials (AEPs), elicited by continuous and stable auditory stimuli, require subjects to respond to various types of audio stimuli during data collection and represent a stable paradigm suitable for brain fingerprint identification. For 80 Hz auditory stimuli, Seha et al. [34] represented individual-specific features using AEPs' energy distribution across different frequency bands. They achieved a single-session identification rate of 96.46% with an error rate as low as 0% for 40 subjects. The cross-session average identification rate ranged from 94.5 to 96.5%, with error rates between 2 and 4%.

In brain fingerprint identification endogenous spontaneous tasks, Miao et al. proposed a multi-loss domain adapter identity recognition method for emotional EEG [35]. They utilized deep neural networks to extract deep features from EEG data and employed domain adaptation to mitigate the temporal variability of EEG data and the influence of emotions on brainprint features. The cross-time individual recognition accuracy reached 94.8%, and it achieved 100% for cross-emotional individual recognition. Additionally, to enhance the robustness of identity recognition systems, Cheng et al. proposed a hybrid authentication method [36] that combines the EEG and eye movement data features. In the scenario of user password input, the average cross-time classification accuracy can reach 88.07%.

However, existing multi-session brain fingerprint identification methods overlook the data distribution differences caused by temporal variability of EEG signals at different sessions, making it challenging for models to extract stable brainprint features across sessions.

1.2.3 Brain Fingerprint Identification for Multi-task EEG

The above studies demonstrate the effectiveness of brain fingerprint identification based on specific mental tasks, but they still have certain limitations. These methods rely on the active cooperation of the subjects, while their performance in mental tasks may fluctuate across different sessions. This variability can make it challenging to ensure the robustness of identification systems in practical applications, as the unstable states of the subjects can easily influence them.

To address these issues, Puengdang et al. [37] combined the SSVEP and ERP paradigms and utilized a Long Short-Term Memory (LSTM) recurrent neural network for brainprint feature extraction and identity prediction. This approach achieved an accuracy of 91.44%, demonstrating the effectiveness of dual mental task brain fingerprint identification. Kong et al. also researched brain fingerprint identification involving specific sessions and multiple mental tasks. They proposed algorithms for brainprint feature extraction, such as amplitude-phase augmentation and low-rank matrix decomposition. These methods were applied to four mental tasks, including resting state, emotion induction, motor imagery, and actual grasping movements, achieving a recognition accuracy of 99.98% for 15 subjects [38, 39]. Furthermore, they introduced deep learning models such as residual multi-scale spatio-temporal convolutional neural networks for extracting task-independent brainprint features, achieving a recognition accuracy of 99.88% [40]. Additionally, Wang et al. [41] developed a brain fingerprint identification method suited for various mental task scenarios. By calculating phase-locking values to estimate functional connectivity within and between frequency bands, EEG signals are transformed into graph representations. Using graph convolutional neural networks, they automatically extracted deep, intrinsic identity features from these EEG signals. When applied to paradigms such as resting state, attention tasks, and image description tasks, their method achieved an average recognition accuracy of 98.94% for 59 subjects in the same state and 98.96% for subjects in different states, demonstrating that EEG signals retain strong identity discriminability even across different mental states.

While the brain fingerprint identification technologies for EEG data collected across multiple mental tasks have made strides in extracting task-independent brainprint features, several challenges persist. EEG data can be affected by individual physiological conditions and often suffers from sample imbalance. This imbalance makes it challenging for data-driven deep neural network methods to extract brainprint features from individuals with few samples. Consequently, models trained on imbalanced datasets may exhibit bias toward the classes with more samples, leading to distorted feature representations.

1.2.4 Brain Fingerprint Identification for Multi-session and Multi-task EEG

In real-world applications, brain fingerprint identification must often span multiple sessions and accommodate a variety of mental tasks. As a result, the system must consistently and accurately identify individuals, regardless of when the data is collected or the mental task involved, ensuring robustness and reliability. Currently, Arnau-González et al. [42] recently introduced a Biometric EEG Dataset for brain fingerprint identification. Collected over three separate sessions, the dataset includes EEG recordings from 21 subjects performing four different mental tasks: rest, emotion induction, mathematical operations, and visual stimulation. Huang et al. [43] established a Multi-subject, Multi-session, and Multi-task Database to investigate EEG commonality and variability. This dataset includes EEG recordings from 106 subjects, with 95 subjects participating in two experiments conducted on different days. The entire experiment comprised six paradigms: rest, transient sensation, steady-state sensation, classic cognition, motor execution, and steady-state sensation with selective attention, involving 14 types of EEG signals. Kumar et al. [44] released a multi-task EEG dataset that recorded EEG signals from 30 subjects engaged in various mental tasks. Each subject participated in data collection experiments at least twice and up to five times at different intervals.

To validate the feasibility of brain fingerprint identification across multiple sessions and mental tasks, Maiorana and Campisi [45] conducted experiments on a dataset comprising 45 subjects. The EEG signals of the subjects were collected over three years, spanning five to six different sessions and utilizing four distinct paradigms. They extensively evaluated the impact of individual aging on the ability to discern identity from EEG signals, considering factors such as data collection paradigms, subject count, session count, and session duration. The study showed that although aging affects EEG features, the impact of a one-month interval on identification performance is minimal, whereas more pronounced effects are observed over extended sessions. Through statistical analysis and performance evaluation using different EEG features and Hidden Markov Models as classifiers, identification error rates below 2% were achieved. Subsequently, the team proposed models based on CNN, RNN [29] and Siamese CNN [46]. They conducted experimental validation on an EEG dataset from 45 subjects across five sessions, each performing six distinct tasks. The experimental results demonstrated that, when considering all six collection protocols collectively for training, regardless of the specific mental task being performed, the identification rates achieved through single-frame recognition could exceed 90%. The proposed deep learning methods outperformed traditional classifiers and existing deep learning networks in biometric recognition tasks based on EEG signals. In a study involving five auditory stimulation tasks, Vinothkumar et al. [47] used features such as autoregressive coefficients and power spectral density. They implemented various classifiers, including the Universal Background Model-Gaussian Mixture Model (UBM-GMM). The results indicated that, by using power spectral density features from all frequency bands, the single-session UBM-GMM achieved recognition accuracy of 89.5% with 70 s of EEG signals, while the cross-session recognition accuracy was 78.6%. This performance surpassed that of other machine learning algorithms, such as k-nearest neighbors and artificial neural networks. Additionally, Kumar et al. [48] initially proposed a subspace technique to enhance biometric information and suppress other interfering factors, effectively projecting individual features of multi-channel EEG into a subspace. Subsequently, they fused this subspace technique with the UBM-GMM model to propose a task-independent brain fingerprint identification method, which achieved a cross-session recognition accuracy of 86.4% in validating multi-session, multi-scenario brain fingerprint identification across 30 subjects performing 12 different task scenarios [44].

Despite notable advancements, brain fingerprint identification continues to encounter significant challenges, primarily due to the intricate interplay between external factors across different sessions and internal factors associated with subjects' fluctuating cognitive states. Most existing studies on cross-task and cross-session brain fingerprint identification predominantly focus on assessing the influence of varying data collection sessions and changes in subjects' mental tasks on recognition performance. However, there is a pronounced lack of comprehensive research aimed at addressing these challenges, leaving the field in its early stages of development. Therefore, further investigation is essential to understand the effects of temporal variability in EEG data and the uncertainties surrounding subjects' cognitive states on brain fingerprint identification. This understanding is crucial to ensure the system operates stably and reliably under complex conditions.

1.2.5 Summary of Related Work

The preceding subsections offer a detailed overview of various categories of brain fingerprint identification. In general, brainprint features derived from EEG signals

1.2 Related Work

have attracted significant attention and made considerable progress in recent years, thanks to their advantages, including non-stealability, continuous identification, anti-spoofing detection, and resistance to coercion. However, EEG signals are characterized by a low signal-to-noise ratio and time-varying properties, making them susceptible to interference by individual cognitive states and external environmental factors. Furthermore, the costs and time constraints associated with data acquisition often lead to limited cognitive tasks available for analysis. Consequently, brain fingerprint identification continues to face numerous challenges that urgently require further in-depth research. The major challenges that persist in existing brain fingerprint identification are summarized below:

- **Difficulty in extracting stable brainprint features across sessions**. As EEG signals are characterized by low signal-to-noise ratio, time-varying, etc., and are easily affected by changes in the external environment and the state of the subjects, cross-session brain fingerprint identification is very challenging. However, most of the current studies focus on specific sessions, and there are relatively few studies on cross-session brain fingerprint identification. In addition, there are large distributional differences between EEG data collected in different sessions, leading to the variability of brainprint features extracted by deep neural networks based on the assumption of data homogeneous distribution under different data distributions, i.e., the problem of cross-session feature instability. Therefore, it is necessary to study how to accurately extract brainprint features that are stable and invariant across sessions to improve the cross-temporal robustness of brain fingerprint identification.
- **Lack of research on task-independent brain fingerprint identification**. Most existing brain fingerprint identification methods require subjects to perform specific tasks, often necessitating external stimuli or active cooperation in particular thought processes or imagery for successful identification. This reliance on task-specific conditions poses significant limitations for real-world applications. Therefore, it is essential to explore ways to overcome the constraints of EEG cognitive tasks, thereby enhancing the practicality of brain pattern recognition. Additionally, EEG signals reflect various aspects of an individual's identity, emotional state, cognitive condition, environmental influences, and session variability. However, there are often pseudo-correlations between identity-related features and non-identity-related features. Consequently, the correlation information learned by models during the training phase may not apply to EEG signals from tasks conducted in unknown sessions or mental states, ultimately diminishing the model's generalization ability and versatility.

In conclusion, this book aims to introduce a range of brain fingerprint identification methods tailored for specific-task multi-session EEG data, multi-task specific-session EEG data, and multi-task multi-session EEG data. These methods are designed to address the challenges of cross-session instability and the limitations of cognitive tasks in existing brain fingerprint identification approaches.

References

1. Tolosana R, Vera-Rodriguez R, Fierrez J, Morales A, Ortega-Garcia J (2020) DeepFakes and beyond: a survey of face manipulation and fake detection. Inform Fusion 64:131–148
2. Mirsky Y, Lee W (2021) The creation and detection of DeepFakes: a survey. ACM Comput Surv (CSUR) 54(1):1–41
3. Armstrong BC, Ruiz-Blondet MV, Khalifian N, Kurtz KJ, Jin Z, Laszlo S (2015) Brainprint: assessing the uniqueness, collectability, and permanence of a novel method for ERP biometrics. Neurocomputing 166:59–67
4. Da Silva J, Castanheira HD, Perez O, Misic B, Baillet S (2021) Brief segments of neurophysiological activity enable individual differentiation. Nat Commun 12(1):5713
5. Van De Ville D, Farouj Y, Preti MG, Liégeois R, Amico E (2021) When makes you unique: temporality of the human brain fingerprint. Sci Adv 7(42):0751
6. Shenghao W, Ramdas A, Wehbe L (2022) Brainprints: identifying individuals from magnetoencephalograms. Commun Biol 5(1):852–863
7. Gui Q, Ruiz-Blondet MV, Laszlo S, Jin Z (2019) A survey on brain biometrics. ACM Comput Surv (CSUR) 51(6):1–38
8. Zhang S, Yang W, Mou H, Pei Z, Li F, Xia W (2024) An overview of brain fingerprint identification based on various neuroimaging technologies. IEEE Trans Cogn Dev Syst 16(1):151–164
9. Zhang S, Sun L, Mao X, Cuiyun H, Liu P (2021) Review on EEG-based authentication technology. Comput Intell Neurosci 1–20:2021
10. La Rocca D, Campisi P, Vegso B, Cserti P, Kozmann G, Babiloni F, De Vico Fallani F (2014) Human brain distinctiveness based on EEG spectral coherence connectivity. IEEE Trans Biomed Eng 61(9):2406–2412
11. Fraschini M, Hillebrand A, Demuru M, Didaci L, Marcialis GL (2014) An EEG-based biometric system using eigenvector centrality in resting state brain networks. IEEE Signal Process Lett 22(6):666–670
12. Thomas KP, Prasad Vinod A (2018) EEG-based biometric authentication using gamma band power during rest state. Circ Syst Signal Process 37:277–289
13. Luis Alfredo Moctezuma and Marta Molinas (2020) Towards a minimal EEG channel array for a biometric system using resting-state and a genetic algorithm for channel selection. Sci Rep 10(1):14917
14. Ma L, Minett JW, Blu T, Wang WSY (2015) Resting state EEG-based biometrics for individual identification using convolutional neural networks. In: International conference of the IEEE engineering in medicine and biology society (EMBC). IEEE, pp 2848–2851
15. Koike-Akino T, Mahajan R, Marks TK, Wang Y, Watanabe S, Tuzel O, Orlik P (2016) High-accuracy user identification using EEG biometrics. In: International conference of the ieee engineering in medicine and biology society (EMBC). IEEE, pp 854–858
16. Rathi N, Singla R, Tiwari S (2021) A novel approach for designing authentication system using a picture based P300 speller. Cogn Neurodyn 805–824
17. Zhang Y, Shen H, Li M, Dewen H (2023) Brain biometrics of steady state visual evoked potential functional networks. IEEE Trans Cogn Dev Syst 15(4):1694–1701
18. Chen Y, Atnafu AD, Schlattner I, Weldtsadik WT, Roh M-C, Kim HJ, Lee S-W, Blankertz B, Fazli S (2016) A high-security EEG-based login system with RSVP stimuli and dry electrodes. IEEE Trans Inform Forensics Secur 11(12):2635–2647
19. Zhang R, Zeng Y, Tong L, Shu J, Runnan L, Li Z, Yang K, Yan B (2022) EEG identity authentication in multi-domain features: a multi-scale 3D-CNN approach. Front Neurorobot 16:901765
20. Nguyen P, Tran D, Huang X, Sharma D (2012) A proposed feature extraction method for EEG-based person identification. In: Proceedings of the international conference on artificial intelligence (ICAI), pp 1–6

References

21. Wu B, Meng W, Chiu W-Y (2022) Towards enhanced EEG-based authentication with motor imagery brain-computer interface. In: Proceedings of the computer security applications conference, pp 799–812
22. Bidgoly AJ, Bidgoly HJ, Arezoumand Z (2022) Towards a universal and privacy preserving EEG-based authentication system. Sci Rep 12(1):2531
23. Moctezuma LA, Molinas M (2018) EEG-based subjects identification based on biometrics of imagined speech using EMD. In: Brain informatics. Springer, pp 458–467
24. Wilaiprasitporn T, Ditthapron A, Matchaparn K, Tongbuasirilai T, Banluesombatkul N, Chuangsuwanich E (2019) Affective EEG-based person identification using the deep learning approach. IEEE Trans Cogn Dev Syst 12(3):486–496
25. Li W, Yi Y, Wang M, Peng B, Zhu J, Song A (2022) A novel tensorial scheme for EEG-based person identification. IEEE Trans Instrum Meas 72:1–17
26. La Rocca D, Campisi P, Scarano G (2013) On the repeatability of EEG features in a biometric recognition framework using a resting state protocol. In: Proceedings of the international conference on bio-inspired systems and signal processing. SciTePress, pp 419–428
27. Ma MK-H, Lee T, Fong MC-M, Wang WS (2020) Resting-state EEG-based biometrics with signals features extracted by multivariate empirical mode decomposition. In: IEEE international conference on acoustics, speech and signal processing (ICASSP). IEEE, pp 991–995
28. Nakamura T, Goverdovsky V, Mandic DP (2017) In-ear EEG biometrics for feasible and readily collectable real-world person authentication. IEEE Trans Inform Forensics Secur 13(3):648–661
29. Maiorana E (2020) Deep learning for EEG-based biometric recognition. Neurocomputing 410:374–386
30. Piciucco E, Maiorana E, Falzon O, Camilleri KP, Campisi P (2017) Steady-state visual evoked potentials for EEG-based biometric identification. In: International conference of the biometrics special interest group (BIOSIG). IEEE, pp 1–5
31. Zhang Y, Li M, Shen H, Hu D (2021) A permanency investigation of SSVEP signals in brain biometrics. In: International conference on electronics technology (ICET). IEEE, pp 758–762
32. Zeng Y, Qunjian W, Yang K, Tong L, Yan B, Shu J, Yao D (2018) EEG-based identity authentication framework using face rapid serial visual presentation with optimized channels. Sensors 19(1):6
33. Zhao H, Wang Y, Liu Z, Pei W, Chen H (2019) Individual identification based on code-modulated visual-evoked potentials. IEEE Trans Inform Forensics Secur 14(12):3206–3216
34. Seha SNA, Hatzinakos D (2020) EEG-based human recognition using steady-state AEPs and subject-unique spatial filters. IEEE Trans Inform Forensics Secur 15:3901–3910
35. Miao Y, Jiang W, Nuo S, Shan J, Jiang T, Zuo N (2023) MLDA: Multi-loss domain adaptor for cross-session and cross-emotion EEG-based individual identification. IEEE J Biomed Health Inform 27(12):5767–5778
36. Cheng S, Wang J, Sheng D, Chen Y (2023) Identification with your mind: a hybrid BCI-based authentication approach for anti-shoulder-surfing attacks using EEG and eye movement data. IEEE Trans Instrum Meas 72:1–14
37. Puengdang S, Tuarob S, Sattabongkot T, Sakboonyarat B (2019) EEG-based person authentication method using deep learning with visual stimulation. In: International conference on knowledge and smart technology (KST). IEEE, pp 6–10
38. Kong W, Jiang B, Fan Q, Zhu L, Wei X (2018) Personal identification based on brain networks of EEG signals. Int J Appl Math Comput Sci 28(4):745–757
39. Kong X, Kong W, Fan Q, Zhao Q, Cichocki A (2018) Task-independent EEG identification via low-rank matrix decomposition. In: IEEE international conference on bioinformatics and biomedicine (BIBM). IEEE, pp 412–419
40. Zhu Y, Peng Y, Song Y, Ozawa K, Kong W (2021) RAMST-CNN: a residual and multi-scale spatio-temporal convolution neural network for personal identification with EEG. IEICE Trans Fundam Electron, Commun Comput Sci 104(2):563–571
41. Wang M, El-Fiqi H, Hu J, Abbass HA (2019) Convolutional neural networks using dynamic functional connectivity for EEG-based person identification in diverse human states. IEEE Trans Inform Forensics Secur 14(12):3259–3272

42. Arnau-González P, Katsigiannis S, Arevalillo-Herráez M, Ramzan N (2021) BED: a new data set for EEG-based biometrics. IEEE Internet Things J 8(15):12219–12230
43. Huang G, Zhenxing H, Chen W, Zhang S, Liang Z, Li L, Zhang L, Zhang Z (2022) M3CV: a multi-subject, multi-session, and multi-task database for EEG-based biometrics challenge. NeuroImage 264:119666
44. Kumar MG, Narayanan S, Sur M, Murthy HA (2021) Evidence of task-independent person-specific signatures in EEG using subspace techniques. IEEE Trans Inform Forensics Secur 16:2856–2871
45. Maiorana E, Campisi P (2017) Longitudinal evaluation of EEG-based biometric recognition. IEEE Trans Inform Forensics Secur 13(5):1123–1138
46. Maiorana E (2021) Learning deep features for task-independent EEG-based biometric verification. Pattern Recogn Lett 143:122–129
47. Vinothkumar D, Kumar MG, Kumar A, Gupta H, Saranya M, Sur M, Murthy HA (2018) Task-independent EEG based subject identification using auditory stimulus. In: Proceeding of workshop on speech, music and mind, issue 2018, pp 26–30
48. Kumar MG, Saranya MS, Narayanan S, Sur M, Murthy HA (2019) Subspace techniques for task-independent EEG person identification. In: International conference of the IEEE engineering in medicine and biology society (EMBC). IEEE, pp 4545–4548

Open Access This chapter is licensed under the terms of the Creative Commons Attribution-NonCommercial-NoDerivatives 4.0 International License (http://creativecommons.org/licenses/by-nc-nd/4.0/), which permits any noncommercial use, sharing, distribution and reproduction in any medium or format, as long as you give appropriate credit to the original author(s) and the source, provide a link to the Creative Commons license and indicate if you modified the licensed material. You do not have permission under this license to share adapted material derived from this chapter or parts of it.

The images or other third party material in this chapter are included in the chapter's Creative Commons license, unless indicated otherwise in a credit line to the material. If material is not included in the chapter's Creative Commons license and your intended use is not permitted by statutory regulation or exceeds the permitted use, you will need to obtain permission directly from the copyright holder.

Chapter 2
Basics of EEG Signals

Abstract This chapter provides a foundational understanding of electroencephalography (EEG) signals, focusing on their generation, acquisition, and the datasets relevant to brain fingerprint identification. The chapter begins with a detailed discussion of how EEG signals are generated through the electrical activity of neurons in the brain. It explains the mechanisms underlying neuronal communication and the resultant electrical potentials that can be captured at the scalp using electrodes. This section establishes the biological basis for EEG signal acquisition and highlights the significance of signal characteristics in the context of identity recognition. Following the generation of EEG signals, the chapter addresses the acquisition process, outlining the technical aspects of EEG recording, including electrode placement, signal amplification, and noise reduction techniques. It emphasizes the importance of proper acquisition methods to ensure high-quality data for subsequent analysis. The chapter concludes with an overview of EEG datasets specifically curated for brain fingerprint identification. This section discusses the criteria for dataset selection, the diversity of stimuli used in experiments, and the relevance of these datasets for training and validating brain fingerprint identification algorithms. By providing a comprehensive understanding of EEG signals and their application in brain fingerprint identification, this chapter lays the groundwork for the advanced methodologies and analyses explored in the subsequent chapters of the book.

2.1 Generation of EEG Signals

Electroencephalogram (EEG) signals are generated through the electrical activity of neurons in the brain [1]. Neurons communicate via synaptic transmissions, creating small electrical impulses that propagate through the brain tissue. When large groups of neurons fire in synchrony, these electrical potentials can be detected at the scalp using electrodes. EEG signals are recorded as voltage fluctuations over time, typically ranging from 0.5 to 100 Hz in frequency. The signal is influenced by various brain activities such as cognition, motor function, and emotions, making EEG a valuable tool for studying brain function and disorders.

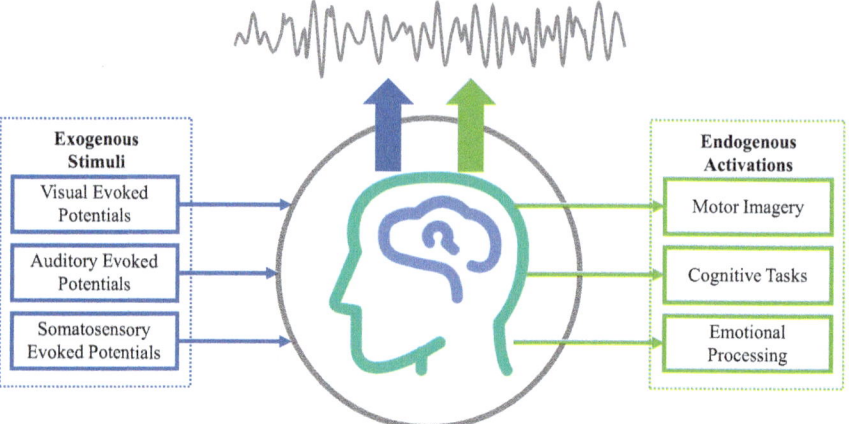

Fig. 2.1 EEG signals based on different paradigms

Researchers typically utilize standardized experimental designs or stimulation protocols to collect EEG data to examine the brain's electrical activity under various conditions. These procedures, known as paradigms, are structured to explore specific cognitive, perceptual, or motor processes. EEG signal generation can be broadly classified into two categories based on the paradigm: exogenous stimuli and endogenous activations, as shown in Fig. 2.1. Exogenous stimuli involve the reception of external environmental inputs, such as visual, auditory, or tactile information, which trigger corresponding nervous system responses and physiological electrical activity. In contrast, endogenous activations do not depend on external stimuli. They arise from the subject's engagement in cognitive tasks, such as motor imagery, emotional responses, attention focus, or other mental activities, leading to spontaneous EEG signals.

2.1.1 EEG Signals Based on Exogenous Stimuli

Common exogenous stimuli include visual, auditory, and tactile stimuli. Different types of stimuli can elicit corresponding evoked potentials (EPs) in subjects.

2.1.1.1 Visual Stimulus-Based Evoked EEG Signals

Visual stimulus-based evoked EEG signals, also known as visually evoked potentials (VEPs), are electrical responses generated in the brain in reaction to visual stimuli. When the eyes are exposed to visual input—such as light flashes, patterns, or images, the brain's visual cortex processes this information, producing detectable changes in the EEG signal. These evoked potentials are time-locked to the onset of the stimulus and typically appear as distinct waveforms within milliseconds. VEPs are commonly used in research and clinical diagnostics to assess the functionality of the visual pathways, detect visual impairments, and study neural processing related to visual perception. Steady-state visually evoked potentials (SSVEPs) [2] and Code-modulated visually evoked potentials (cVEPs) [3] are both types of evoked EEG signals based on visual stimuli, but their generation mechanisms and characteristics differ.

SSVEPs are EEG signals elicited when the brain is exposed to repetitive visual stimuli, such as flickering lights or oscillating visual patterns at a fixed frequency. These stimuli cause a continuous and periodic response in the visual cortex that matches the frequency of the stimulus, allowing SSVEPs to be easily detected in the EEG recordings. SSVEPs are commonly used in brain-computer interfaces (BCIs) and neurofeedback systems due to their reliability and high signal-to-noise ratio. SSVEPs are also widely used to study attention, visual processing, and non-invasive communication in clinical settings. Their frequency-specific characteristics make them an invaluable tool for real-time monitoring of brain responses to visual stimuli.

cVEPs are EEG signals generated in response to visual stimuli modulated by a pseudorandom or deterministic code. Instead of using simple, periodic visual stimuli like SSVEPs, cVEPs rely on more complex patterns of visual flashes or flickers, encoded using binary sequences (e.g., m-sequences or Gold codes). These codes are designed to have specific mathematical properties that make the evoked potentials more distinguishable from background EEG noise. The distinct advantage of cVEPs lies in their ability to offer robust, high-rate information transfer with reduced susceptibility to noise and interference compared to traditional VEPs. As a result, they are increasingly employed in BCIs for fast and reliable communication, enabling users to select targets based on modulated visual stimuli. Additionally, cVEPs are used in vision research to investigate neural encoding and decoding mechanisms.

2.1.1.2 Auditory Stimulus-Based Evoked EEG Signals

Auditory evoked potentials (AEPs) [4] are electrical responses generated in the brain following auditory stimuli, such as brief tones or sound pulses, and complex sequences, such as musical compositions or segments of speech. These responses reflect the brain's processing of auditory information.

In AEPs, the evoked potentials produced by the brain reflect the transmission and processing of auditory stimuli through the auditory pathways. Typically, AEPs include waveform components such as N1, P2, and N2, which reflect the brain's perception, encoding, and processing of auditory stimuli. Common auditory stimulus paradigms include binaural hearing paradigms [5] and attention-based auditory paradigms [6]. AEPs are widely used in clinical settings for hearing assessments, including the diagnosis of hearing loss and auditory processing disorders. They are also employed in research to study auditory perception, cognitive processing, and the neural mechanisms underlying sound discrimination and localization. Research has shown that brain fingerprint identification based on auditory evoked potentials exhibits good stability in cross-session scenarios.

2.1.1.3 Event-Related Potentials

Event-related potentials (ERPs) [7] are EEG activities generated by subjects in response to processing various stimulus events. ERPs are time-locked and phase-locked to the occurrence of the stimulus, allowing for analysis based on the timing of the event. Typically, within a few hundred milliseconds after the event, the brain generates specific electrical activities that can be recorded and associated with the event.

In brain-computer interfaces, the P300 ERP has a long history of development. This potential is a large positive slow wave generated approximately 250–750 ms after the subject receives an oddball stimulus. The oddball paradigm is used to study the brain's neural response to unexpected or unusual events. In an oddball experiment, subjects are required to continuously receive a series of stimuli, most of which are frequently occurring "standard stimuli", with occasional "oddball stimuli" or "target stimuli". For example, in a visual oddball experiment, the standard stimulus might be a fixed pattern or shape, while the target stimulus might be a different pattern or shape. Subjects are required to focus on and record the appearance of the target stimulus, typically measuring reaction time or accuracy.

In an auditory oddball experiment, the standard stimulus might be a series of repetitive sounds or tones, while the target stimulus might be a different sound or tone. Subjects need to respond when they hear the target stimulus. Currently, ERP-based brain fingerprint identification has been extensively studied.

ERPs offer excellent temporal resolution, enabling the tracking of brain activity on a millisecond timescale. They are widely used in cognitive neuroscience, psychology, and clinical research to explore processes such as language comprehension, memory, and attention. Additionally, ERPs are valuable in clinical settings for assessing neurological disorders and cognitive impairments, providing insights into the timing and nature of cognitive processes in the human brain. Meanwhile, existing studies have developed effective brain fingerprint identification techniques based on ERPs.

2.1.2 EEG Signals Based on Endogenous Stimulation

EEG signals based on endogenous stimulation include physiological electrical signals generated by the subject's voluntary motor imagery, emotions, or other mental activities.

2.1.2.1 Motor Imagery-Based EEG Signals

These EEG signals record the brain's electrical activity during motor imagery, where subjects are asked to imagine performing specific movements, such as grasping with the left or right hand or raising an arm, without actually performing the movements. When individuals imagine performing a specific motor task, distinct electrical activity is generated in regions of the brain associated with motor control, particularly in the sensorimotor cortex. These imagined movements produce measurable changes in the EEG signal, often characterized by event-related desynchronization (ERD) [8] and event-related synchronization (ERS) in specific frequency bands, such as alpha (8–12 Hz) and beta (12–30 Hz) rhythms. The analysis of these signals can provide insights into the neural mechanisms underlying motor planning and execution.

Motor imagery-based EEG signals have significant applications in BCIs, rehabilitation therapies, and cognitive neuroscience research. They enable users to control devices or prosthetics through thought alone and can aid in recovery from motor impairments by facilitating neuroplastic changes in the brain. Additionally, studying these signals helps researchers understand the cognitive processes involved in motor planning and the brain's representation of movement. Currently, motor imagery-based endogenous EEG signals have gained prominence in brain fingerprint identification research due to their convenience and applicability across various scenarios without requiring external stimuli, making them highly suitable for individual identification.

2.1.2.2 Affective EEG Signals

Affective EEG signals refer to the brain's electrical activity that is associated with emotional processing and responses. These signals are captured through EEG and reveal how the brain reacts to various emotional stimuli, such as images, sounds, or memories. Clinical research has shown that theta (3–8 Hz) and alpha (8–12 Hz) bands are associated with cognitive or memory load [9, 10], and asymmetry in alpha band activity in the prefrontal cortex during resting state correlates with the severity of depression [11]. Based on these physiological phenomena and EEG signal characteristics, researchers can analyze EEG signals using signal processing techniques and deep learning algorithms to extract features related to emotions and cognition, thereby detecting an individual's internal state.

In current research on affective EEG, inducing different emotional states in subjects presents a significant challenge. To capture EEG signals corresponding to various emotional states, researchers typically employ two primary methods of emotion induction. The first method is subjective induction, where subjects are instructed to express different emotions through facial expressions or recall specific emotional scenarios to elicit the desired feelings. The second method is event induction, where subjects are presented with specific images, sounds, or videos designed to evoke emotional responses. These methods of emotion induction provide researchers with effective means to obtain affective EEG signals, thereby facilitating the exploration of the relationship between emotions and EEG activity.

Affective EEG signals have a wide range of applications, including mental health assessments, affective computing, and marketing research. They provide valuable insights into the neural correlates of emotions and deepen our understanding of human emotional experiences. In the realm of brain fingerprint identification, researchers are increasingly utilizing EEG signals to investigate how various emotional states impact individual identity recognition. By incorporating emotional state information, they aim to enhance the stability and robustness of brain fingerprinting methods.

2.2 Acquisition of EEG Signals

To obtain high-quality EEG data, a complete EEG acquisition system includes electrodes, amplifiers with filters, analog-to-digital converters, and a data recording computer. Specifically, the voltage signals from electrodes on the scalp are converted into appropriate voltage ranges by amplifiers and then converted into digital format by analog-to-digital converters for storage on the data recording computer.

In most EEG research, EEG signals are recorded simultaneously from multiple electrodes distributed across the scalp. This multi-channel electrode acquisition enables researchers to decompose the collected data into different components, such as artifacts, thereby enhancing the accuracy of EEG feature extraction. High-density electrode arrays can indeed capture EEG data with greater spatial resolution, providing more detailed information. However, this approach also entails higher equipment

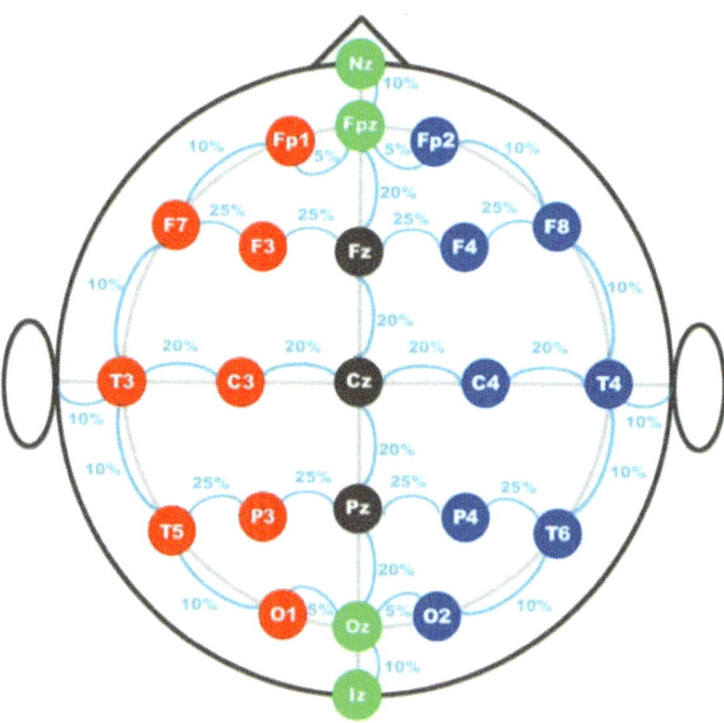

Fig. 2.2 The electrode distribution according to the International 10–20 system

2.2 Acquisition of EEG Signals

Table 2.1 International 10–20 standard lead electrode and cerebral cortex

Cerebral region	Electrode name
Frontal pole	Fp
Frontal	F
Central	C
Parietal	P
Occipital	O
Temporal	T

costs and more time-consuming data processing and analysis. Additionally, increased electrode channels can introduce more noise due to impedance between the scalp and electrodes, potentially compromising the quality of EEG data. To ensure the comparability of results, current EEG acquisition devices adopt the 10–20 electrode placement system published by the International Federation of Societies for Electroencephalography and Clinical Neurophysiology, commonly referred to as the International 10–20 system. This system defines four reference points: the nasion, the inion, and the left and right preauricular points. Other electrodes are positioned at intervals of 10 or 20%, as illustrated in Fig. 2.2. This layout aligns electrode positions with the anatomical relationship of the cerebral cortex. The letters in Fig. 2.2 correspond to different brain regions, as shown in Table 2.1. The numbers in the electrode labels indicate the hemisphere and distance from the midline, with odd numbers representing the left hemisphere and even numbers representing the right hemisphere from the subject's perspective. Larger numbers denote greater distances from the midline, while electrodes positioned at the midline are labeled with "z", such as Fz, Cz, and Oz.

The acquired EEG signals are physiological signals characterized by high randomness, weak amplitudes, and susceptibility to irrelevant noise, resulting in various artifacts such as eye movement artifacts, muscle artifacts, and sweat artifacts. Consequently, EEG signals directly recorded from scalp electrodes often fail to reflect the neural signals from the brain accurately. To improve data quality, preprocessing, and denoising the acquired raw EEG data are essential to minimize or eliminate the influence of these artifacts. Effectively removing artifacts from EEG signals while preserving the original EEG information is crucial for subsequent analysis and research applications of EEG data.

Although standardized experimental paradigms and operating procedures are adhered to during the generation and data acquisition of EEG signals, EEG signals recorded from the same subject in different sessions still exhibit non-stationarity. This phenomenon may arise from the combined effects of various factors. First, changes in individual physiological states at other sessions, including wakefulness, fatigue level, and emotional state, can influence brain electrical activity and thus the characteristics of EEG signals. Second, variations in external environmental factors, including light, noise, and temperature, may disrupt brain electrical activity, thereby impacting the acquisition and characteristics of EEG signals. Additionally, uncertainties in experimental conditions, such as inconsistencies in experimental operations, may further compromise the stability of EEG signals. The cross-session non-stationarity of EEG signals implies differences in the distribution of EEG data across different sessions, leading to significant performance degradation when models trained on EEG data collected during one period are applied to data from other periods. This variability challenges the practical application and further development of EEG-based brainwave recognition technology in real-world scenarios.

2.3 EEG Dataset for Brain Fingerprint Identification

This section details EEG datasets related to brain fingerprint identification and categorizes them according to the mental tasks and sessions.

2.3.1 Specific-Task and Specific-Session EEG Dataset

- **Motor Imagery Dataset** [12]
  The motor imagery dataset, publicly available from the BNCI Horizon 2020 project, contains EEG recordings from 14 participants performing a two-class motor imagery task. Based on the cue-guided Graz BCI training paradigm [13], the session included eight runs: five for training and three for feedback validation. Here, the five training runs are used as experimental data. Each run comprises 20 trials, with participants mentally simulating right hand or foot movement for 5 s per trial. A biosignal amplifier was used to record EEG signals, with data sampled at 512 Hz.
- **Neuromarketing Dataset (NMK)** [14]
  The NMK dataset is gathered from an experiment involving 20 healthy participants, equally divided by gender and with no history of neurological issues. The stimulus includes a 10-min video composed of an 8-min neutral segment interspersed with six 30-s advertising clips. During recording, subjects sit quietly in a controlled environment, focusing on the screen ahead and passively viewing the video with limited knowledge of the experiment's aims. The EEG data are collected using a 16-channel G-Tec device with a 256 Hz sampling rate.
- **Fatigue Driving Dataset (DRI)** [15]
  The DRI dataset is recorded from a fatigue driving experiment with 12 participants, aged 23–25, who were all licensed drivers, right-hand dominant, and free from neurological or psychiatric conditions. A driving simulator with a replica cab is employed, presenting participants with 8 different conditions. In each scenario, subjects focused on maintaining control of the vehicle. EEG signals are recorded via a 16-channel G-Tec device at a sampling rate of 256 Hz, with electrode impedance maintained below 5 kΩ.

2.3 EEG Dataset for Brain Fingerprint Identification

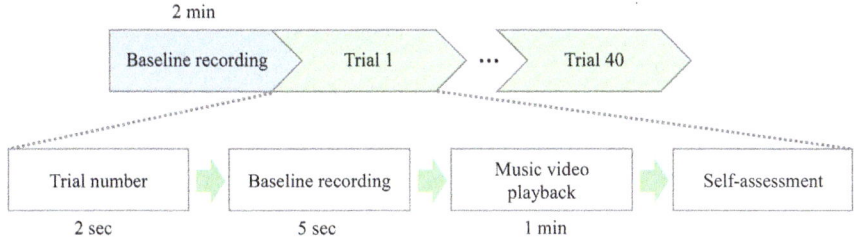

Fig. 2.3 The protocol of DEAP dataset

- **DEAP Dataset** [16]
  DEAP dataset is a multimodal dataset for the analysis of human affective states. The EEG and peripheral physiological signals of 32 participants are recorded as each watches 40 one-minute long excerpts of music videos. Participants rate each video in terms of the levels of arousal, valence, like/dislike, dominance, and familiarity. For 22 of the 32 participants, a frontal face video is also recorded. A novel method for stimuli selection is used, utilizing retrieval by affective tags from the last.fm website, video highlight detection, and an online assessment tool. The protocol of the EEG acquisition is shown in Fig. 2.3.
- **EEG Motor Movement/Imagery Dataset (EEGMMI)** [17]
  The dataset used the 64-channel acquisition equipment corresponding to the BCI2000 system [18] to collect the electroencephalogram of 109 subjects. It is according to the two baseline runs about resting state (one with eyes open, one with eyes closed) and four tasks (open and close left or right fist, imagine opening and closing left or right fist, open and close both fists or feet and imagine opening and closing both fists or feet) suggested by the BCI experimental paradigm. The first experimental run of every task contained is used for analysis. The sampling rate was 160 Hz. To explore the recognition ability of the model under different tasks, the dataset is divided into EEGMMI-RS with resting states (experimental runs 1, 2) and EEGMMI-MI with four tasks (experimental runs 3–6). The protocol of the EEG acquisition is shown in Fig. 2.4. In total, the dataset contains 109 subjects. The data of 4 subjects was damaged after data inspection, hence, their recording was not further considered.

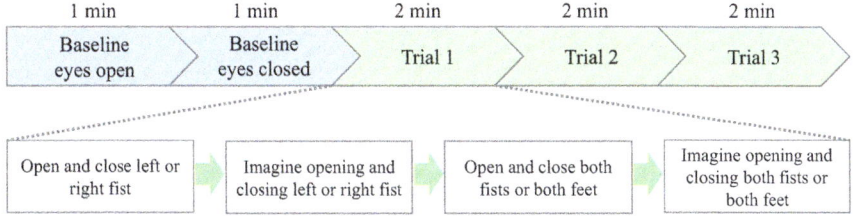

Fig. 2.4 The protocol of EEG Motor Movement/Imagery Dataset

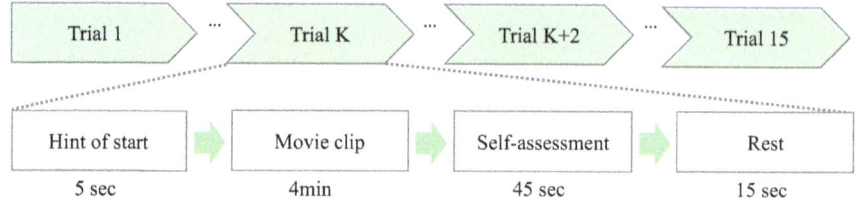

Fig. 2.5 The protocol of SJTU emotion EEG Dataset

2.3.2 Specific-Task and Multi-session EEG Dataset

- **SJTU Emotion EEG Dataset (SEED)** [19]
  The SEED dataset is a significant resource for researchers in the field of BCI and affective computing, developed by Shanghai Jiao Tong University. This dataset comprises EEG data from 15 subjects who participated in multiple sessions using stimulus materials. EEG signals are recorded using a 62-electrode cap following the international 10–20 electrode placement system and a sampling rate of 1000 Hz. On average, the time interval between sessions is approximately 19.4 d. In each session, participants watched movie clips representing three broad emotional categories: positive, neutral, and negative, featuring 5 clips per category. The protocol of the EEG acquisition is shown in Fig. 2.5. The sequence and content of the clips remained consistent across sessions.
- **SJTU Emotion EEG Dataset IV (SEED-IV)** [20]
  The SEED-IV dataset is an extension of the original SEED dataset, designed to enhance research in emotion recognition and BCI applications. This dataset recording experiments included a total of 15 healthy subjects (eight females) aged between 20 and 24 years. The subjects are required to watch the film clips that are carefully selected to induce different types of emotion including happy, sad, fearful, and neutral. For each subject, 3 sessions are performed on different days, and each session contained 24 trials. The protocol of the EEG acquisition is shown in Fig. 2.6. The EEG signals are collected with the 62-channel ESI NeuroScan System at a sampling rate of 1000 Hz and downsampled to a 200 Hz sampling rate.

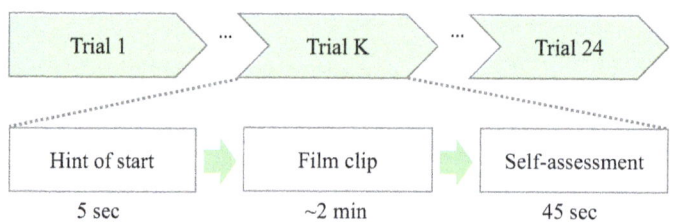

Fig. 2.6 The protocol of SJTU emotion EEG Dataset IV

2.3 EEG Dataset for Brain Fingerprint Identification

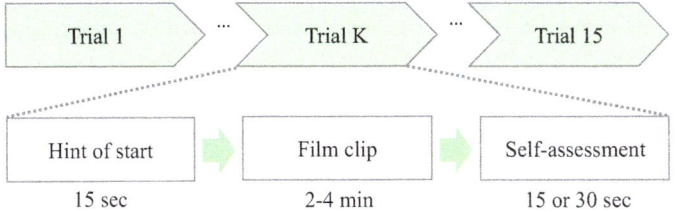

Fig. 2.7 The protocol of SJTU emotion EEG Dataset V

- **SJTU Emotion EEG Dataset V (SEED-V)** [21]
  The SEED-V dataset is the latest iteration in the series of emotion-related EEG datasets developed by Shanghai Jiao Tong University, aimed at advancing research in affective computing. This dataset comprises EEG data from 16 subjects collected over three sessions. It encompasses five more detailed emotional categories: happy, sad, disgust, neutral, and fear, each emotion includes three movie clips, resulting in a total of 15 clips. The protocol of the EEG acquisition is shown in Fig. 2.7. The maximum time gap between sessions is restricted to 20 d, with an average interval of approximately 8.1 d. In each session, clips of diverse emotional categories are presented in alternating order. Nevertheless, the content and sequence of clips differed across the three sessions.
- **Cross-Session Dataset for Collaborative RSVP-Based BCI (RSVP-Based Dataset)** [22]
  The dataset focuses on collaborative BCI research, featuring simultaneous EEG recordings from two subjects performing synchronized target detection tasks. It includes data from 14 participants across two sessions, recorded on different days with an average interval of approximately 23 d. Whole-head EEG data are collected using 62 channels, with raw recordings provided for flexibility in analysis. The experimental design involved subjects identifying target images of humans within sequences of street images presented at a rate of 10 Hz, comprising three blocks of 14 trials, each containing 100 images with 4 target images. In total, the dataset encompasses 84 blocks (1176 trials) and allows for precise synchronization of event-triggered epochs, making it especially useful for studying cross-session event-related potential (ERP) detection algorithms in both individual and collaborative RSVP-based BCI systems.
- **BCI Graz Dataset A** [23]
  The dataset consists of EEG recordings from 9 subjects participating in a cue-based brain-computer interface (BCI) paradigm that includes four different motor imagery tasks: imagining movements of the left hand (class 1), right hand (class 2), both feet (class 3), and the tongue (class 4). Each subject underwent two recording sessions on different days, with each session comprising 6 runs separated by short breaks. Each run consisted of 48 trials, with 12 trials for each of the four classes, resulting in a total of 288 trials per session. The protocol of the EEG acquisition is

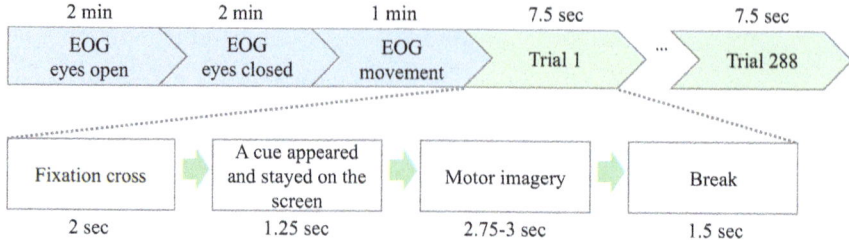

Fig. 2.8 The protocol of BCI Graz dataset A

shown in Fig. 2.8. EEG data were collected using 22 Ag/AgCl electrodes, arranged with inter-electrode distances of 3.5 cm. The montage followed a specific layout, with all signals recorded monopolarly, utilizing the left mastoid as a reference and the right mastoid as ground. The recordings were sampled at a rate of 250 Hz and bandpass-filtered between 0.5 and 100 Hz, with an amplifier sensitivity set to 100 μV. Additionally, a 50 Hz notch filter was applied to suppress line noise, ensuring high-quality data acquisition for subsequent analysis.

- **P300 EEG Dataset** [24]

  This dataset focuses on event-related potentials, specifically the P300 component, associated with attention and stimulus evaluation-related cognitive processes. The experimental protocol consisted of two sessions performed on two different days. The first is a BCI session: ALS participants are asked to control a 6 by 6 P300 speller. During the second session, the participants are involved in the screening of their attentional filtering efficiency and working memory capacity and are asked to perform two behavioral tasks: the rapid serial visual presentation task and the change detection task. Scalp EEG signals are recorded (g.MOBILAB, g.tec, Austria) from eight channels according to 10–10 standard (Fz, Cz, Pz, Oz, P3, P4, PO7, and PO8) using active electrodes (g.Ladybird, g.tec, Austria). All channels are referenced to the right earlobe and grounded to the left mastoid. The EEG signal is digitized at 256 Hz. Data acquisition and stimuli delivery are managed by the BCI2000 framework.

- **RSVP-Based Dataset** [22]

  This RSVP dataset is designed for collaborative Brain-Computer Interface (BCI) research and includes EEG data from 14 subjects, divided into seven groups. Each group consisted of two subjects performing the same target image detection task synchronously. All subjects participated in two sessions, with an average interval of approximately 23 d between sessions. The dataset aims to evaluate cross-session BCI performance, and the evaluation results indicate that appropriate signal processing algorithms can significantly improve BCI performance in both individual and collaborative conditions. Collaborative BCI methods, which fuse brain activity from multiple subjects, show significantly enhanced performance compared to

2.3 EEG Dataset for Brain Fingerprint Identification

Table 2.2 Description of 128-channel Multi-task EEG dataset

No.	Mental task	Brief description
1	Resting state	In this state, when the word "OPEN" appears on the screen, the participant is instructed to open their eyes and avoid blinking. Conversely, when the word "CLOSE" is displayed, the participant is required to close their eyes and refrain from excessive eye movements
2	Watching movie clips	The subjects are asked to view two distinct video clips. The first clip is a dramatic battle scene from Avengers: Age of Ultron, followed by a serene, peaceful documentary from Home. This contrast is intended to evoke differing cognitive and emotional responses
3	Hand movement imagination	In this task, participants are instructed to imagine performing the hand movements shown in a video. The video displays 25 repetitions of gripping motions for both the left and right hand, and the participant must mentally simulate the corresponding actions
4	Hand-grip movement	A video showcasing either the left or right-hand gripping motion is presented randomly. The participant is then asked to perform the corresponding physical action, with 25 repetitions of each hand's gripping movement shown

individual BCI systems. The dataset was recorded using a 64-channel EEG system with a sampling rate of 500 Hz. This dataset provides valuable resources for developing more efficient cross-session BCI algorithms, advancing the performance and practicality of collaborative RSVP-based BCI systems.

2.3.3 Multi-task and Specific-Session EEG Dataset

- **Multi-Task EEG Dataset (MTED) [25]**
  The dataset used in this study was self-collected and reviewed, receiving approval from the Institutional Ethical Review Committee at Saitama Institute of Technology (Protocol No. 2018-01). EEG signals were recorded from 15 subjects, aged 20–25 years, all of whom were free from any neurological disorders. The data were collected following the international 10–20 system using a 62-channel ESI NeuroScan system. The original sampling rate of 1000 Hz was downsampled to 200 Hz for analysis. The experiment was conducted in a single session, during which participants performed a series of mental tasks, including resting state, emotion induction via movie clips, motor imagery, and hand-grip movements. Detailed descriptions of each mental task can be found in Table 2.2. The EEG acquisition protocol is illustrated in Fig. 2.9.

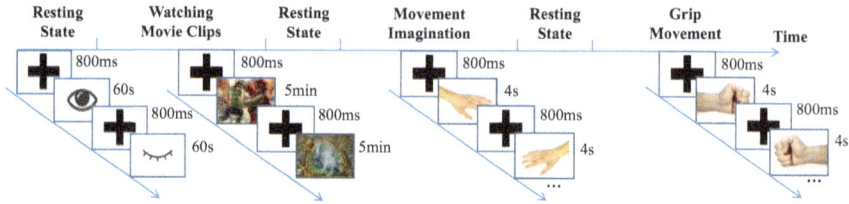

Fig. 2.9 The protocol of multi-task EEG dataset

2.3.4 Multi-task and Multi-session EEG Dataset

- **128-channel Multi-task EEG Dataset (128-MTED)** [26]
  This dataset consists of the EEG data collected from 30 subjects with 12 elicitation protocols, including Odd Ball Classic, Odd Ball Stereo, Imagining Binary Answer, Semantically Opposite Words, Familiar and Unfamiliar words, Proper Improper, Motor and Mental Imaginary, Passive Audio, Passive Audio Stereo, Odd Ball Visual, SSVEP, and Passive Audio-Visual. The first nine tasks are conducted with closed eye condition, and the last three tasks are conducted with open eye condition. Detailed descriptions of the mental tasks are provided in Table 2.3. Each subject performs at least 2 and at most 5 sessions, the subjects randomly perform no less than 4 different tasks in a single session. The average time between training and the last testing session is 44 d. It is worth noting that each subject did not perform all the tasks. The EEG data were collected using a 128-channel dense-array EEG system manufactured by Electrical Geodesics, Inc (EGI) at a sampling rate of 250 Hz.

2.3 EEG Dataset for Brain Fingerprint Identification

Table 2.3 Description of 128-channel Multi-task EEG dataset

No.	Mental task	Brief description	Participants
1	Odd Ball Classic	Participants were presented with frequent non-target stimuli and infrequent target stimuli. The target and non-target stimuli consisted of audio beeps that differed in frequency or duration	13
2	Odd Ball Stereo	Similar to No. 1, the target and non-target stimuli consisted of audio beeps played in the left and right ears	12
3	Imagining Binary Answers	A set of binary questions was presented to participants. They were asked to first imagine the answer and then respond with a mouse click	7
4	Semantically Opposite Words	Semantically opposite words, such as "yes" and "no", were played to the subject over multiple trials. The subject was instructed to respond with left and right mouse clicks depending on the semantics of the word being played	4
5	Familiar and Unfamiliar Words	The subjects were presented with common and uncommon words. They were instructed to respond with a mouse click upon hearing a familiar word	6
6	Proper and Improper Sentences	Regular and ill-formed sentences were presented to the subject, who was required to respond with a mouse click upon hearing ill-formed sentences	8
7	Motor and Mental Imaginary	Participants were asked to imagine motor movements, such as rotating their left and right fists. For the mental imagery task, they were instructed to count numbers in reverse	6
8	Passive Audio	Participants passively listened to a variety of audio stimuli, including music, sentences, stories, and attention-triggering sounds, such as sirens	17
9	Passive Audio Stereo	Similar to No. 7, the auditory stimuli were played through only one ear (either left or right) at a time using headphones	11
10	Odd Ball Visual	Similar to No. 1, the target and non-target stimuli consist of visual objects varying in shape and color	6
11	Steady-State Visually Evoked Potential	Visual objects flickering at different frequencies were displayed to participants. At the end of each trial, a question about the shape or color of the object was asked	12
12	Passive Audio-Visual	Audio-visual clips were played to the participants, and at the end of each clip, a question based on the stimuli was asked	12

References

1. Niedermeyer E, Lopes da Silva FH (2005) Electroencephalography: basic principles, clinical applications, and related fields. Lippincott Williams & Wilkins
2. Liu Q, Chen K, Ai Q, Xie SQ (2014) Recent development of signal processing algorithms for SSVEP-based brain computer interfaces. J Med Biol Eng 34(4):299–309
3. Martínez-Cagigal V, Thielen J, Santamaria-Vazquez E, Pérez-Velasco S, Desain P, Hornero R (2021) Brain-computer interfaces based on code-modulated visual evoked potentials (c-VEP): a literature review. J Neural Eng 18(6):061002
4. Picton TW, Hillyard SA, Krausz HI, Galambos R (1974) Human auditory evoked potentials I: evaluation of components. Electroencephalogr Clin Neurophysiol 36:179–190
5. Jeremy Hill N, Lal TN, Bierig K, Birbaumer N, Scholkopf B (2004) Attention modulation of auditory event-related potentials in a brain-computer interface. In: IEEE international workshop on biomedical circuits and systems. IEEE, pp 3–5
6. Guo J, Gao S, Hong B (2010) An auditory brain-computer interface using active mental response. IEEE Trans Neural Syst Rehabil Eng 18(3):230–235
7. Picton TW et al (1992) The P300 wave of the human event-related potential. J Clin Neurophysiol 9:456–456
8. Pfurtscheller G, Da Silva Lopes FH (1999) Event-related EEG/MEG synchronization and desynchronization: basic principles. Clin Neurophysiol 110(11):1842–1857
9. Diaz-Piedra C, Victoria Sebastián M, Di Stasi LL (2020) EEG theta power activity reflects workload among army combat drivers: an experimental study. Brain Sci 10(4):199
10. Brouwer A-M, Hogervorst MA, Van Erp JBF, Heffelaar T, Zimmerman PH, Oostenveld R (2012) Estimating workload using EEG spectral power and ERPs in the n-back task. J Neural Eng 9(4):045008
11. Miljevic A, Bailey NW, Murphy OW, Prabhavi M, Perera N, Fitzgerald PB (2023) Alterations in EEG functional connectivity in individuals with depression: a systematic review. J Affect Disord 328:287–302
12. Steyrl D, Scherer R, Faller J, Müller-Putz GR (2016) Random forests in non-invasive sensorimotor rhythm brain-computer interfaces: a practical and convenient non-linear classifier. Biomed Eng/Biomedizinische Technik 61(1):77–86
13. Pfurtscheller G, Neuper C (2001) Motor imagery and direct brain-computer communication. Proc IEEE 89(7):1123–1134
14. Kong W, Zhao X, Hu S, Vecchiato G, Babiloni F (2013) Electronic evaluation for video commercials by impression index. Cogn Neurodyn 7(6):531–535
15. Kong W, Lin W, Babiloni F, Sanqing H, Borghini G (2015) Investigating driver fatigue versus alertness using the granger causality network. Sensors 15(8):19181–19198
16. Koelstra S, Muhl C, Soleymani M, Lee J-S, Yazdani A, Ebrahimi T, Pun T, Nijholt A, Patras I (2011) Deap: a database for emotion analysis; using physiological signals. IEEE Trans Affect Comput 3(1):18–31
17. Goldberger AL, Amaral LAN, Glass L, Hausdorff JM, Ivanov PC, Mark RG, Mietus JE, Moody GB, Peng C-K, Eugene Stanley H (2000) PhysioBank, PhysioToolkit, and PhysioNet: components of a new research resource for complex physiologic signals. Circulation 101(23):e215–e220
18. Schalk G, McFarland DJ, Hinterberger T, Birbaumer N, Wolpaw JR (2004) BCI2000: a general-purpose brain-computer interface (BCI) system. IEEE Trans Biomed Eng 51(6):1034–1043
19. Zheng W-L, Bao-Liang L (2015) Investigating critical frequency bands and channels for EEG-based emotion recognition with deep neural networks. IEEE Trans Auton Mental Dev 7(3):162–175
20. Zheng W-L, Liu W, Yifei L, Bao-Liang L, Cichocki A (2018) Emotionmeter: a multimodal framework for recognizing human emotions. IEEE Trans Cybern 49(3):1110–1122
21. Liu W, Qiu J-L, Zheng W-L, Lu B-L (2021) Comparing recognition performance and robustness of multimodal deep learning models for multimodal emotion recognition. IEEE Trans Cogn Dev Syst 14(2):715–729

References

22. Zheng L, Sun S, Pei W, Gao X, Zhang L, Wang Y (2020) A cross-session dataset for collaborative brain-computer interfaces based on rapid serial visual presentation. Front Neurosci 14:579469
23. Brunner C, Leeb R, Müller-Putz G, Schlögl A, Pfurtscheller G (2008) BCI competition 2008—Graz data set A
24. Riccio A, Simione L, Schettini F, Pizzimenti A, Inghilleri M, Belardinelli MO, Mattia D, Cincotti F (2013) Attention and p300-based BCI performance in people with amyotrophic lateral sclerosis. Front Hum Neurosci 7(1):732
25. Kong X, Kong W, Fan Q, Zhao Q, Cichocki A (2018) Task-independent EEG identification via low-rank matrix decomposition. In: 2018 IEEE international conference on bioinformatics and biomedicine (BIBM). IEEE, pp 412–419
26. Kumar MG, Narayanan S, Sur M, Murthy HA (2021) Evidence of task-independent person-specific signatures in EEG using subspace techniques. IEEE Trans Inform Forensics Secur 16:2856–2871

Open Access This chapter is licensed under the terms of the Creative Commons Attribution-NonCommercial-NoDerivatives 4.0 International License (http://creativecommons.org/licenses/by-nc-nd/4.0/), which permits any noncommercial use, sharing, distribution and reproduction in any medium or format, as long as you give appropriate credit to the original author(s) and the source, provide a link to the Creative Commons license and indicate if you modified the licensed material. You do not have permission under this license to share adapted material derived from this chapter or parts of it.

The images or other third party material in this chapter are included in the chapter's Creative Commons license, unless indicated otherwise in a credit line to the material. If material is not included in the chapter's Creative Commons license and your intended use is not permitted by statutory regulation or exceeds the permitted use, you will need to obtain permission directly from the copyright holder.

Chapter 3
Specific-Task and Multi-session Brain Fingerprint Identification with Multi-scale Graph Neural Network

Abstract Electroencephalogram (EEG) signals have emerged as a promising biometric modality due to their inherent potential for secure personal identification. However, a persistent challenge in EEG-based biometric systems is the influence of affective state variations, which can compromise the reliability of data acquisition regardless of the protocols employed. Moreover, the non-stationary nature of EEG further exacerbates this issue, as fluctuations in affective states over time can introduce significant variability into the signal. As such, achieving accurate brain fingerprint identification in the presence of affective state changes remains a critical hurdle. This chapter introduces the Multi-scale Convolution and Graph Pooling Network (MCGP), a novel model designed to address these challenges. The MCGP framework employs multiple 1D convolutional layers operating at different scales to dynamically extract and integrate relevant features from EEG signals. Additionally, the model incorporates a graph pooling layer with an attention mechanism, which facilitates the generation of hierarchical graph embeddings. These embeddings are subsequently concatenated and passed through a fully connected classification layer for final identification. Experiments on the SEED and SEED-V datasets show that MCGP achieves an average accuracy of 85.51% on SEED and 88.69% on SEED-V under cross-session conditions with mixed affective states. When a single affective state is maintained across sessions, MCGP reaches 85.75% accuracy on SEED and 88.06% on SEED-V for the same affective state, while achieving 79.57 and 84.52% for different affective states, respectively. These results demonstrate that MCGP effectively mitigates the impact of affective state variations, outperforming baseline methods. Notably, identification performance is slightly better for sessions with the same affective state than for those with different affective states.

3.1 Introduction

Biometric identification refers to the automatic recognition of individuals based on their physiological or behavioral characteristics [1]. EEG signals, which capture electrical potentials generated by neural activity in the brain's cerebral cortex, are acquired non-invasively using specialized scalp-mounted devices. Compared to traditional biometric modalities such as fingerprints, facial recognition, and voice recognition [2–4], EEG-based biometrics present greater challenges in terms of forgery and offer enhanced confidentiality. Additionally, EEG-based identification provides an extra layer of security through the possibility of revocation, achieved by modifying the stimuli used for elicitation [5].

Despite considerable advancements in EEG-based brain fingerprint identification, a critical issue remains unresolved: the non-stationary nature of EEG data [6] makes it vulnerable to fluctuations in a subject's affective states. Regardless of the protocol used to collect EEG data, these affective states can significantly influence identification performance [7]. Several studies have investigated this phenomenon. Wang et al. [8] explored the impact of different affective states by integrating functional connectivity and graph convolutional neural networks (GCNN). Wilaiprasitporn et al. [9] applied a cascaded deep learning approach to evaluate individual identification performance under varying affective states. Similarly, Arnau et al. [7] examined the influence of affective states. However, most studies have focused on the effects within a single session, which does not align with real-world scenarios that require multiple-session identification.

To address these challenges, we propose an integrated Multi-scale Convolution and Graph Pooling network (MCGP) to mitigate the influence of affective state variations across different sessions in brain fingerprint identification. Our experimental setup uses two publicly available datasets: SEED, which includes three coarse-grained affective states (negative, neutral, and positive), and SEED-V, which features five more detailed affective states (fear, disgust, neutral, and happy). Each dataset consists of EEG data collected across three separate sessions.

The MCGP workflow begins with dynamic EEG feature extraction through multi-scale convolution. These extracted features are then used to construct a graph, which serves as input for the graph pooling layer. After passing through this layer, the features are transformed into graph embedding representations using an attention mechanism. Finally, identity recognition is performed using fully connected layers.

Our contributions, when compared to existing studies, are as follows:

- MCGP introduces a novel EEG feature extraction framework and reduces the impact of affective state variations across sessions on brain fingerprint identification.
- At a detailed level, we explore cross-session scenarios, treating each affective state separately as either a training or test set. Our results show that identification performance is slightly better for sessions with the same affective state compared to sessions with different affective states.

3.2 Multi-scale Convolution and Graph Pooling Network

As shown in Fig. 3.1, the MCGP framework consists of three key modules, each designed to address specific challenges in EEG-based brain fingerprint identification. This section provides a detailed overview of these components.

3.2.1 Dynamic Feature Extraction Module

This module consists of multi-scale convolutional layers and an attention fusion layer, with the primary objective of dynamically extracting features at different scales for each channel and performing feature fusion.

The multi-scale convolutional layers use a set of 1D convolutional kernels with varying scales. The kernel size for the kth level of the convolutional layer is determined by the ratio coefficient $\alpha_k \in \mathbb{R}$ and the length of the sample, denoted as Q. The parameter k ranges from 1 to K. In this chapter, we set α_k to [0.1, 0.2, 0.5] with $K = 3$. Consequently, the scale of the convolutional kernel for the kth level, denoted as S^k, is calculated as follows:

$$S^k = (1, \alpha_k \cdot Q). \tag{3.1}$$

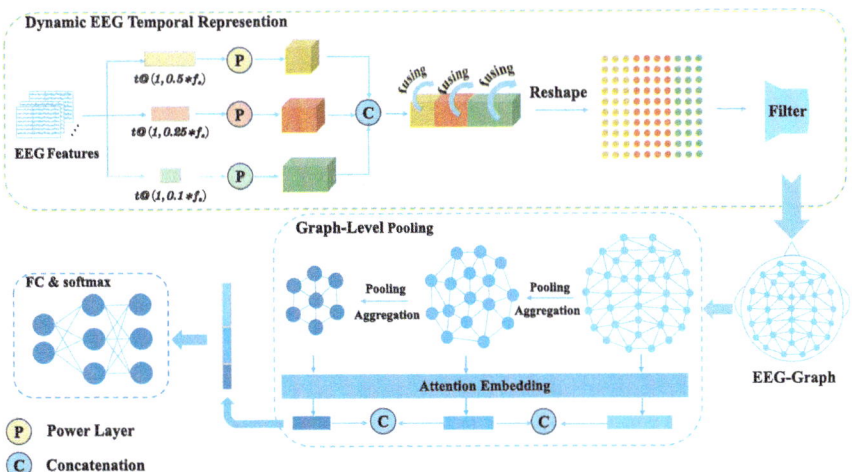

Fig. 3.1 The framework of the Multi-scale Convolution and Graph Pooling network (MCGP). The process begins with the preprocessing of EEG data, followed by the application of multi-scale 1D convolution layers. The resulting outputs, each corresponding to different feature sets, are represented in the figure with distinct colors. These features are then fused and used to construct the EEG graph. In the graph-level pooling layer, a cluster soft assignment matrix probabilistically assigns nodes in the EEG graph to distinct clusters. Finally, the graph embedding vectors from each layer are concatenated and passed through a fully connected layer for classification

The input to the multi-scale convolutional layers is the preprocessed EEG data, denoted as $X_i \in \mathbb{R}^{C \times Q}$, where $i \in [1, \ldots, N]$. Here, N represents the number of EEG data samples, and C is the number of channels in the input EEG data. The output of the kth level of convolution is denoted as $X_{T\_out}^k \in \mathbb{R}^{t \times C \times f_k}$, where t represents the number of convolution kernels and f_k denotes the length of the output features. The formulation for $X_{T\_out}^k$ is as follows:

$$X_{T\_out}^k = \Phi_{\log}\left(\mathcal{F}_{AP}\left(\Phi_{\text{square}}\left(\mathcal{F}_{\text{Conv2D}}\left(X_i, S^k\right)\right)\right)\right), \quad (3.2)$$

where $\mathcal{F}_{\text{Conv2D}}(X_i, S^k)$ denotes the convolutional operation applied to X_i using a kernel of size S^k. The function $\Phi_{\text{square}}(\cdot)$ represents squaring, while $\mathcal{F}_{AP}(\cdot)$ denotes average pooling, and $\Phi_{\log}(\cdot)$ is the logarithmic transformation. For the SEED and SEED-V datasets, the pooling size and stride are set to (1, 4) and (1, 1), respectively. This approach enables the dynamic representation of EEG data for each channel [10].

The outputs of the convolutional kernels from all levels are concatenated. For $X_i \in \mathbb{R}^{C \times Q}$, the output of the multi-scale convolutional layers, denoted as $X_{\text{cat}}^i \in \mathbb{R}^{t \times C \times \sum f_k}$, is computed as:

$$X_{\text{cat}}^i = \Gamma\left(X_{T\_out}^1, \ldots, X_{T\_out}^K\right). \quad (3.3)$$

Following concatenation, attention fusion is performed using a 1×1 convolutional layer. The attention-fused representation $\overline{X}_{\text{fuse}}^i \in \mathbb{R}^{t \times C \times 0.5 \cdot \sum f_k}$ for each X_{cat}^i is calculated as:

$$\overline{X}_{\text{fuse}}^i = \mathcal{F}_{bn-AP}\left(\Phi_{\text{Leaky-ReLU}}\left(\mathcal{F}_{\text{fuse}}\left(X_{\text{cat}}^i\right)\right)\right), \quad (3.4)$$

where $\mathcal{F}_{bn-AP}(\cdot)$ represents batch normalization followed by average pooling with window size (1, 2) and stride (1, 2). $\mathcal{F}_{\text{fuse}}(\cdot)$ denotes the 1×1 convolution, and $\Phi_{\text{Leaky-ReLU}}(\cdot)$ is the Leaky-ReLU activation function.

To reshape $\overline{X}_{\text{fuse}}^i$ into $X_{\text{fuse}}^i \in \mathbb{R}^{C \times t \cdot 0.5 \cdot \sum f_k}$, representing the attributes of each node in the EEG graph, the following operation is applied:

$$X_{\text{fuse}}^i = \mathcal{F}_{\text{reshape}}\left(\overline{X}_{\text{fuse}}^i\right). \quad (3.5)$$

Before entering the next module, $X_{\text{fuse}}^i \in \mathbb{R}^{C \times t \cdot 0.5 \cdot \sum f_k}$ is filtered using a learnable filter matrix $W_{\text{filter}} \in \mathbb{R}^{C \times t \cdot 0.5 \cdot \sum f_k}$. The filter weights operate on each feature representation of each channel individually:

$$X_{\text{filtered}}^i = \Phi_{\text{ReLU}}\left(W_{\text{filter}} \circ X_{\text{fuse}}^i - b_{\text{filter}}\right), \quad (3.6)$$

where $\circ$ denotes the Hadamard product.

3.2.2 Graph-Level Pooling Module

In this module, we employ a hierarchical graph pooling approach by stacking multiple Graph Convolutional Network (GCN) layers [11]. A cluster soft assignment matrix is used to map the outputs from the previous layer, forming clusters and establishing inter-cluster connections. These cluster representations serve as the coarsened inputs for the subsequent GCN layer. The well-established GCN framework [12] is utilized, as represented by the following equation:

$$H^{(l)} = \Phi_{\text{ReLU}} \left(\tilde{D}^{-\frac{1}{2}} \tilde{A} \tilde{D}^{-\frac{1}{2}} H^{(l-1)} W^{(l)} \right), \tag{3.7}$$

In Eq. (3.7), $\tilde{A} = A + I$, where $A \in \mathbb{R}^{C \times C}$ is the adjacency matrix of the graph and I is the identity matrix of the same size as A, adding self-connections to the graph. Furthermore, $\tilde{D} = \sum_j \tilde{A}_{ij}$ denotes the degree of the adjacency matrix, and $W^{(l)} \in \mathbb{R}^{d \times d}$ is the trainable weight matrix. $H^{(l-1)} \in \mathbb{R}^{C \times d}$ represents the node embeddings generated by the previous graph convolutional layer.

In the first graph convolutional layer, we use X_{filtered}^i as the initial input $H^{(0)}$. For the lth layer, the node features $X^{(l)}$ and adjacency matrix $A^{(l)}$ are input into two distinct GCN modules: the Aggregation GCN and the Pooling GCN. These modules generate a new node aggregation representation $X_{\text{agg}}^{(l)} \in \mathbb{R}^{n_l \times d}$ and a cluster soft assignment matrix $S^{(l)} \in \mathbb{R}^{n_l \times n_{l+1}}$, where n_l and n_{l+1} represent the number of nodes (clusters) in layer l and layer $l + 1$, respectively.

The Aggregation GCN is a standard GCN layer, defined as in Eq. (3.7), which takes $A^{(l)}$ and $X^{(l)}$ as inputs:

$$X_{\text{agg}}^{(l)} = \text{GCN}_{l,\text{agg}} \left(A^{(l)}, X^{(l)} \right). \tag{3.8}$$

For the cluster soft assignment matrix $S^{(l)}$, it is obtained using another GCN layer:

$$S^{(l)} = \Phi_{\text{softmax}} \left(\text{GCN}_{l,\text{pool}} \left(A^{(l)}, X^{(l)} \right) \right), \tag{3.9}$$

The softmax function $\Phi_{\text{softmax}}(\cdot)$ is applied row-wise. $S^{(l)}$ is the learned cluster assignment matrix at the lth layer. The number of rows in $S^{(l)}$ is n_l, the number of nodes (clusters) at the lth layer, while the number of columns is n_{l+1}, corresponding to the number of clusters in the next layer. $S^{(l)}$ assigns probabilities to each node in the lth layer for soft assignment to the predefined n_{l+1} clusters.

Next, the goal is to coarsen the adjacency matrix $A^{(l)}$ and the node feature matrix $X_{\text{agg}}^{(l)}$, generating a new adjacency matrix $A^{(l+1)}$ and a pooled feature matrix $X^{(l+1)}$. These will serve as inputs for the $(l + 1)$th layer of the GNN. The following equations describe this process:

$$X^{(l+1)} = S^{(l)^T} X_{\text{agg}}^{(l)} \in \mathbb{R}^{n_{l+1} \times d}, \tag{3.10}$$

$$A^{(l+1)} = S^{(l)^T} A^{(l)} S^{(l)} \in \mathbb{R}^{n_{l+1} \times n_{l+1}}. \tag{3.11}$$

3.2.3 Graph-Level Attention Embedding Module

After obtaining $X_{\text{agg}}^{(l)}$ from each Aggregation GCN layer, the next step is to create a global embedding of the entire graph using an attention mechanism.

Let the cth row of $X_{\text{agg}}^{(l)}$ be denoted as $x_c^{(l)} \in \mathbb{R}^d$, representing the embedding of node c. First, the global context $h_{\text{mean}}^{(l)} \in \mathbb{R}^d$ is computed as the average of the node embeddings:

$$h_{\text{mean}}^{(l)} = \Phi_{\tanh}\left(\left(\frac{1}{n_l}\sum_{c=1}^{n_l} x_c^{(l)}\right) W_m\right), \tag{3.12}$$

where $W_m \in \mathbb{R}^{d \times d}$ is a trainable weight matrix, and $\Phi_{\tanh}(\cdot)$ is the hyperbolic tangent activation function. The global context $h_{\text{mean}}^{(l)}$ captures a combined representation of global structural and feature information.

Using $h_{\text{mean}}^{(l)}$, attention weights are calculated for each node. For each node c, its attention score is determined by taking the inner product with the context vector $h_{\text{mean}}^{(l)}$. This assigns higher attention to nodes that are more similar to the global context [13]. The attention weights are then passed through a softmax function to ensure they lie in the range (0, 1).

Finally, the graph embedding $h_{\text{embed}}^{(l)} \in \mathbb{R}^d$ is computed as a weighted sum of the node embeddings:

$$h_{\text{embed}}^{(l)} = \Phi_{\text{softmax}}\left(h_{\text{mean}}^{(l)} X_{\text{agg}}^{(l)T}\right) X_{\text{agg}}^{(l)}. \tag{3.13}$$

Before being fed into the classifier, the graph embeddings from each layer are concatenated:

$$h_{\text{cat}} = \Gamma\left(h_{\text{embed}}^1, \ldots, h_{\text{embed}}^L\right). \tag{3.14}$$

The classifier is then applied to the concatenated embedding h_{cat} as follows:

$$\hat{y}_i = \Phi_{\text{softmax}}(W h_{\text{cat}} + b), \tag{3.15}$$

where W is a trainable weight matrix, and $b \in \mathbb{R}^{n_{\text{classes}} \times 1}$ represents the bias term.

3.2.4 Optimization

The backpropagation method is used to iteratively optimize the network parameters until the desired criteria are met. The objective function employed is the cross-entropy loss, as given by:

$$L_{CE} = -\sum_{i=1}^{N} y_i \log(\hat{y}_i), \tag{3.16}$$

3.2 Multi-scale Convolution and Graph Pooling Network

The optimization process follows the learning algorithm described in Algorithm 3.1. The main steps include data transformation, feature aggregation, and the application of the Graph-Level Pooling Module to refine node assignments.

Algorithm 3.1 Learning algorithm for MCGP

Input: Training data represented as a matrix $X_i \in \mathbb{R}^{C \times Q}$, where C is the dimensionality of the input data (concatenated differential entropy features across five frequency bands), and Q is the number of time points in each EEG sample. The ground truth label y_i is also required for each sample
Output: Predicted label $\hat{y}_j$ for the test data X_j generated by the MCGP model
1: **Initialization**: Prepare necessary variables and structures for computation
2: **for** all epochs i from 1 to #epochs **do**
3:    **for** k from 1 to 3 **do**
4:      **Determine Temporal Kernel Size**: Calculate the kth temporal kernel size using Eq. (3.1).
5:      **Transform Input Data**: Obtain $X_{T\_out}^{k}$ by applying Eq. (3.2) to X_i
6:    **end for**
7:    **Concatenate Transformed Data**: Create X_i^{cat} using Eq. (3.3)
8:    **Kernel-level Attention Fusion**: Fuse data at the kernel level using Eqs. (3.4) and (3.5) to obtain X_{fuse}^i
9:    **Local Filtering**: Apply local filtering to each node attribute as defined by Eq. (3.6)
10:    **for** l from 1 to L **do**
11:      **Aggregate and Pool Features**: Compute $X_{\text{agg}}^{(l)}$ and $S^{(l)}$ using Eqs. (3.8) and (3.9)
12:      **Update Graph Representation**: Determine $X^{(l+1)}$ and adjacency matrix $A^{(l+1)}$ via Eqs. (3.10) and (3.11)
13:    **end for**
14:    **Graph-level Representation**: Generate h_{cat} using Eqs (3.12)–(3.14)
15:    **Updates**: Update the model using Eq. (3.19)
16: **end for**
17: **Save Model**: Store the trained MCGP model
18: **return Prediction**: Predicted label $\hat{y}_j$ for the given test EEG data X_j

---

Additionally, an auxiliary link prediction objective is introduced to train the Graph-Level Pooling Module. In each layer, our objective is to minimize the following link prediction loss:

$$L_{LP} = \left\| A^{(l)} - S^{(l)} S^{(l)^T} \right\|_F, \quad (3.17)$$

where $\|\cdot\|_F$ denotes the Frobenius norm. The goal here is to improve the structural relationships captured by the adjacency matrix.

Furthermore, we aim to determine the membership of each cluster, suggesting that the output cluster assignment for each node should ideally approximate a one-hot vector. To encourage this, we apply regularization by minimizing the following entropy loss:

$$L_E = \frac{1}{n_l} \sum_{i=1}^{n_l} H(S_i), \qquad (3.18)$$

where $H(\cdot)$ is the entropy function, and S_i refers to the ith row of S.

The overall optimization objective is therefore:

$$\mathcal{L} = \omega_1 L_{CE} + \omega_2 L_{LP} + \omega_3 L_E, \qquad (3.19)$$

where ω_1, ω_2, and ω_3 are hyperparameters. The model parameters are optimized by minimizing this combined loss function. The training procedure of the MCGP model is summarized in Algorithm 3.1.

3.3 Brain Fingerprint Identification with MCGP

This chapter presents two distinct cross-session experimental setups: one that incorporates mixed affective states, where the training set consists of all affective states from a particular session, and another that focuses on single affective states, utilizing data from a specific state across different sessions as the training set.

3.3.1 Data Pre-processing

In this chapter, we utilize two well-established EEG datasets for affective state recognition: SEED [14] and SEED-V [15]. Each dataset undergoes a separate preprocessing pipeline to prepare the data for analysis.

First, the raw EEG data is downsampled to a frequency of 200 Hz, and a bandpass filter is applied to restrict the signal to the frequency range of 1–50 Hz. Next, we compute differential entropy features [16] across five frequency bands, each using a 1-s time window. These bands include: Delta (1–4 Hz), Theta (4–8 Hz), Alpha (8–12 Hz), Beta (12–30 Hz), and Gamma (30–50 Hz). The features extracted from these five frequency bands, across 62 EEG channels, are then concatenated to create a 310-dimensional feature vector for each time window. To ensure consistent input for the model, the data from each session is divided into non-overlapping 10-second segments. For the SEED dataset, each session includes 5085 samples, while the SEED-V dataset has three sessions containing 4384, 3504, and 3872 samples, respectively.

3.3.2 Comparison with Existing Methods

In the experimental section, we benchmark our proposed model against several well-known approaches, including:

- **DGCNN** [17]: A graph convolutional neural network that models each EEG channel as a node and dynamically computes the adjacency relationships between channels using Chebyshev polynomials.
- **EEGNet** [18]: A compact neural network architecture utilizing 1D convolutional filters. It captures spatial relationships through depthwise convolutions and distinguishes within- and between-feature map relationships using separable convolutions.
- **TSception** [19]: A model combining both temporal and spatial convolutional layers. It uses multi-scale 1D convolutional kernels to capture time-frequency representations and exploits brain region asymmetry to enhance global and inter-hemispheric discrimination.
- **LGGNet** [10]: A model that uses varied-length 1D convolutional kernels to dynamically acquire representations, consolidates features from different brain regions, and performs global graph convolution to model complex inter-regional relationships.

These models serve as baselines for our experiments, enabling us to evaluate the performance of our proposed approach and demonstrate its effectiveness in comparison to existing state-of-the-art methods.

3.3.3 Implementation Details

In this section, we outline the implementation details and parameter settings used for training the model. To identify the optimal configuration for our Aggregation and Pooling GCN layers, a grid search was conducted. The hidden layer sizes for both layers were explored within the range [5 : 5 : 100], as determined by our search strategy. We set the maximum number of training epochs to 100 to strike a balance between model performance and training duration. A batch size of 64 was chosen, a typical value that ensures efficient memory usage while maintaining convergence stability. To enhance the model's generalization across both datasets, we applied a dropout rate of 0.5 uniformly.

For optimization, we employed the Adam optimizer, which is well suited for adaptive learning rates. The learning rate was set to 1e-3, based on favorable results observed in our preliminary experiments. All other hyperparameters were held constant to streamline the tuning process and focus on the most significant factors.[1] The

---

[1] For a detailed description of the hyperparameter tuning process and access to the code, please visit: https://github.com/Ultramua/MCGP.git.

implementation was developed using the PyTorch framework, taking advantage of its efficient and flexible deep learning capabilities.

To ensure a fair comparison, the optimal hyperparameters recommended by the authors of the baseline methods were adopted. Moreover, we maintained consistency by utilizing the same training framework and settings for all models, including those of our proposed approach.

In the experimental setup, we conducted cross-session EEG-based identification. The evaluation involved three chronological tasks: "session 1 → session 2", "session 1 → session 3", and "session 2 → session 3". In each task, the training set consisted of samples from the source session, while the test set comprised samples from the target session. For instance, in the "session 1 → session 2" task, the model was trained on data from session 1 and evaluated on its ability to identify individuals in samples from session 2.

3.3.4 Performances for Mixed Affective State Scenario

The accuracies and F1 scores for cross-session EEG-based brain fingerprint identification in mixed affective state scenarios are presented in Tables 3.1 and 3.2. The highest accuracy in each experimental condition is highlighted in bold. The results provide valuable insights, which are summarized as follows:

Based on the experimental results, our proposed model outperforms all other methods in terms of both average accuracy and F1 score across the two datasets. This finding highlights the effective mitigation of the influence of varying affective states on identification performance through the MCGP model.

For example, in the SEED dataset, our model achieved an average accuracy of 85.81%, surpassing EEGNet, DGCNN, TSception, and LGG by 26.57, 22.18, 12.44, and 8.24%, respectively. Notably, our model consistently achieved the highest accuracy in the "session 1 → session 3" task, with a peak accuracy of 90.87%. In terms of the F1 score, our model maintained an average of 85.35%, showing significant

Table 3.1 Cross-session brain fingerprint identification results (%) on SEED dataset for mixed affective state scenarios

Model	Session 1 → Session 2		Session 1 → Session 3		Session 2 → Session 3		Average	
	ACC	F1	ACC	F1	ACC	F1	ACC	F1
EEGNet	50.08	41.23	59.27	50.99	67.46	64.26	58.94	52.16
DGCNN	56.59	50.61	58.28	50.24	75.12	72.55	63.33	57.80
TSception	64.08	60.03	73.97	69.80	81.15	77.13	73.07	68.99
LGG	72.47	67.61	78.28	74.88	81.05	78.73	77.27	73.74
MCGP	**82.18**	**82.12**	**90.87**	**90.43**	**83.47**	**83.50**	**85.51**	**85.35**

3.3 Brain Fingerprint Identification with MCGP

Table 3.2 Cross-session brain fingerprint identification results (%) on SEED-V dataset for mixed affective state scenarios

Model	Session 1 → Session 2		Session 1 → Session 3		Session 2 → Session 3		Average	
	ACC	F1	ACC	F1	ACC	F1	ACC	F1
EEGNet	64.18	56.66	78.36	73.49	77.84	75.37	73.46	68.51
DGCNN	78.67	75.78	80.52	78.06	**90.39**	**90.29**	83.20	81.38
TSception	79.43	75.70	84.24	79.78	82.76	80.22	82.14	78.56
LGG	84.29	81.41	82.50	79.66	88.26	87.97	85.01	83.01
MCGP	**88.40**	**85.96**	**91.07**	**90.88**	86.59	86.50	**88.69**	**87.78**

improvements over EEGNet, DGCNN, TSception, and LGG by 33.19, 27.55, 16.36, and 11.61%, respectively. The highest F1 score of 90.43% was observed in the "session 1 → session 3" task.

While DGCNN achieved a peak accuracy of 90.39% on the SEED-V dataset and performed reasonably well on average, the overall results across both datasets suggest that focusing solely on spatial relationships may not always be optimal for improving performance. However, this does not diminish the role of spatial relationships in EEG signals, as they remain an important factor to some extent. These findings emphasize that modeling spatial relationships after feature extraction is a more effective strategy.

Although our model shares architectural similarities with LGG and TSception–utilizing multi-scale convolution for dynamic EEG feature extraction and spatial relationship modeling–MCGP consistently outperforms these models across various affective states. This outcome underscores the effectiveness of a data-driven approach for learning spatial relationships, offering a promising direction for future research.

3.3.5 Performance in the Single Affective State Scenario

In this scenario, the data from all affective states within a specific session is aggregated for model training. Additionally, the data corresponding to each individual affective state is categorized separately. The training set is composed of data from one specific affective state, while data from a different session is used as the test set. This approach provides a more comprehensive assessment of the model's performance.

The results of the cross-session EEG-based brain fingerprint identification in the single affective state scenario are presented in Figs. 3.2 and 3.3 for the SEED and SEED-V datasets, respectively. In these figures, each row corresponds to a training set derived from a specific affective state, while each column represents a test set from another session. For example, in Fig. 3.2, the first row indicates that data from the negative affective state is used as the training set, with each subsequent column representing the testing set from different sessions.

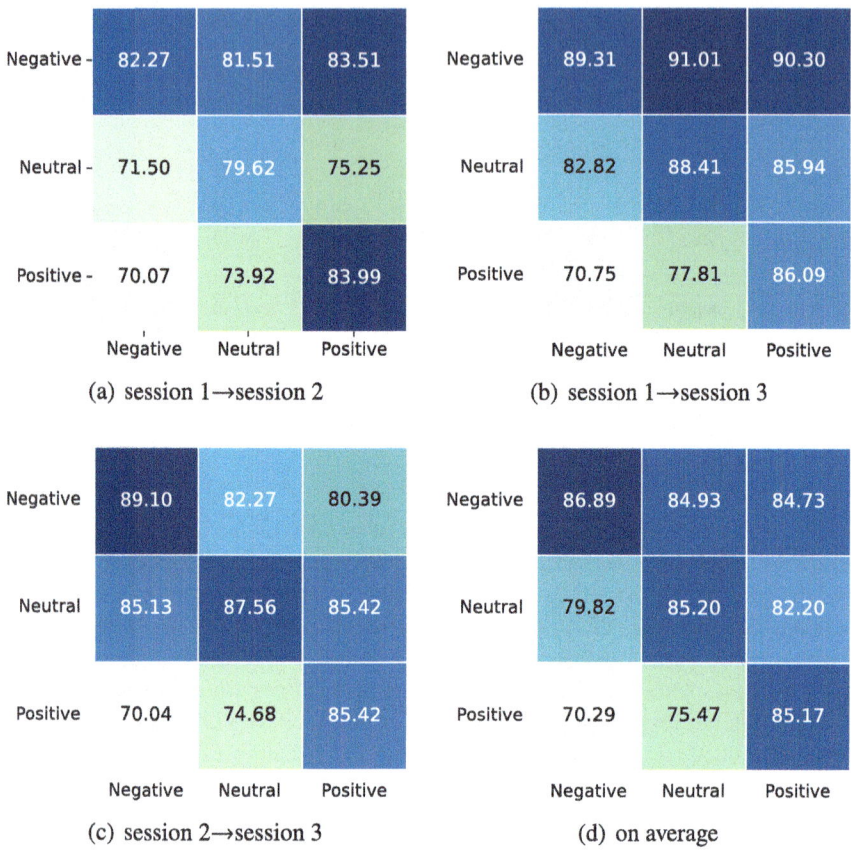

Fig. 3.2 The brain fingerprint identification results (%) for single affective state scenarios in the SEED dataset. Rows represent the affective states of the training data, and columns represent the affective states of the testing data from different sessions

Upon averaging the experimental results, it is clear that the accuracy is higher when the affective state remains the same compared to when different affective states are used. As shown in Fig. 3.4, the recognition accuracy for the same affective state is 6.18% higher on the SEED dataset than for different affective states, while on the SEED-V dataset, the difference is 3.54%. These findings are consistent with the observations reported in Arnau-González et al. [7].

3.3.6 Effect of Each Component

To validate the contribution of each component in our model, we conducted three distinct ablation experiments within the mixed affective state scenario.

3.3 Brain Fingerprint Identification with MCGP

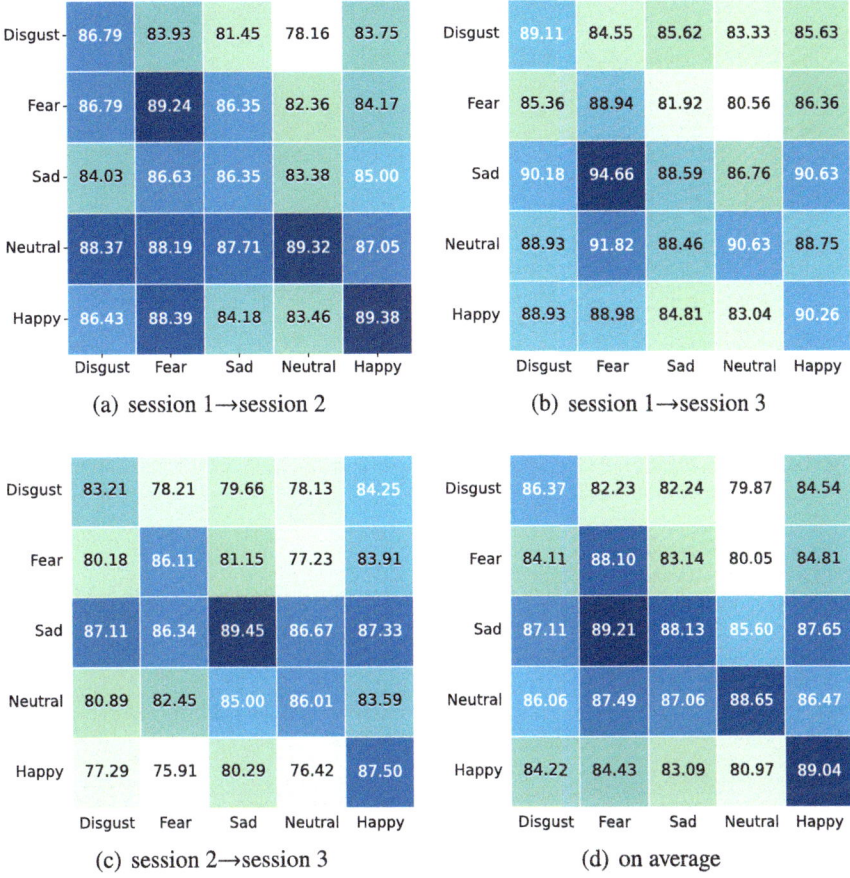

Fig. 3.3 The brain fingerprint identification results (%) for single affective state scenarios in the SEED-V dataset. Rows represent the affective states of the training data, and columns represent the affective states of the testing data from different sessions

Firstly, our proposed method is trained with **Model I**, which utilizes a single-scale convolutional layer. Next, **Model II** is created by replacing the graph pooling layer with a standard graph convolutional network. Finally, **Model III** is constructed by removing the graph-level attention embedding module and replacing it with an embedding vector obtained by averaging.

Table 3.3 provides an overview of the structural compositions of the three models, along with a summary of the performance differences resulting from the modifications made to each component in our proposed framework on both datasets.

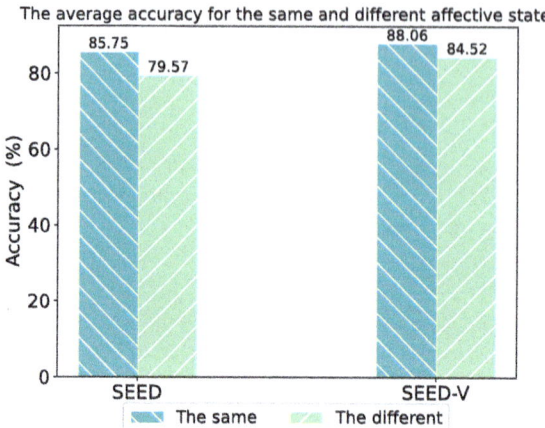

Fig. 3.4 Average identification accuracy for the same and different affective states in the single affective state scenario for SEED and SEED-V datasets

Table 3.3 Model architectures and performance comparison (%) in ablation study on SEED and SEED-V datasets (average of ACC and F1 scores)

Model	Architecture	SEED		SEED-V	
		ACC	F1	ACC	F1
Single-scale convolutional layer					
Model I	Graph-level pooling module	81.50	80.25	85.03	83.68
	Graph-level attention embedding module				
Multi-scale convolutional layer					
Model II	**Standard graph convolutional network**	**71.48**	**65.67**	84.82	83.10
	Graph-level attention embedding module				
Multi-scale convolutional layer					
Model III	Standard graph convolutional network	79.52	73.38	**83.75**	**82.70**
	Average embedding module				
Multi-scale convolutional layer					
MCGP	Graph-level pooling module	85.51	85.35	88.68	87.78
	Graph-level attention embedding module				

The experimental results show that while the use of a single-scale convolutional layer has a relatively minor effect on the model's performance compared to the other components, it still underperforms compared to the full model. In contrast, removing the graph pooling module and the attention embedding mechanism leads to a substantial decline in performance. These findings highlight the critical role of all three components–single-scale convolutional layers, graph pooling, and attention embedding–in maximizing model effectiveness.

3.4 Conclusion

In this chapter, we introduced the Multi-scale Convolution and Graph Pooling Network (MCGP), a novel approach designed to address the challenges posed by affective state variations in cross-session brain fingerprint identification using EEG data. The MCGP model utilizes multi-scale 1D convolutional layers to dynamically extract relevant features from EEG signals, while the graph pooling mechanism learns spatial relationships in a data-driven manner, enhancing the model's robustness.

The proposed method offers several advantages in the context of brain fingerprint identification. Notably, MCGP's ability to mitigate the effects of affective state fluctuations makes it particularly suitable for real-world applications where consistency in affective states cannot always be guaranteed. By leveraging hierarchical feature extraction and graph-based learning, the model effectively enhances identification accuracy under varying conditions, which is critical for the deployment of EEG-based biometric systems.

However, despite its strengths, MCGP also has certain limitations. The performance of the model may still be influenced by extreme variations in affective states, and further refinement is required to improve its robustness in such scenarios. Additionally, the model's reliance on data-driven graph learning may necessitate large and diverse datasets to fully capture the complex relationships within EEG signals.

In summary, the MCGP model presents a promising step forward in the field of EEG-based brain fingerprint identification, offering significant improvements in handling affective state variations. Future research can build upon these findings to further enhance the model's generalizability and applicability to diverse biometric authentication tasks.

References

1. Jain AK, Ross A, Prabhakar S (2004) An introduction to biometric recognition. IEEE Trans Circ Syst Video Technol 14(1):4–20
2. Joshi M, Mazumdar B, Dey S (2023) A novel minutiae-oriented approach for partial fingerprint-based masterprint mitigation. Pattern Recogn. 109935
3. Yin Z, Yiu V, Xiaolin H, Tang L (2021) End-to-end face parsing via interlinked convolutional neural networks. Cogn Neurodyn 15:169–179
4. Vitikainen A-M, Mäkelä E, Lioumis P, Jousmäki V, Mäkelä JP (2015) Accelerometer-based automatic voice onset detection in speech mapping with navigated repetitive transcranial magnetic stimulation. J Neurosci Methods 253:70–77
5. Wang M, Wang S, Jiankun H (2022) Cancellable template design for privacy-preserving EEG biometric authentication systems. IEEE Trans Inform Forensics Secur 17:3350–3364
6. Wan Z, Yang R, Huang M, Zeng N, Liu X (2021) A review on transfer learning in EEG signal analysis. Neurocomputing 421:1–14
7. Arnau-González P, Arevalillo-Herráez M, Katsigiannis S, Ramzan N (2018) On the influence of affect in EEG-based subject identification. IEEE Trans Affect Comput 12(2):391–401
8. Wang M, El-Fiqi H, Hu J, Abbass HA (2019) Convolutional neural networks using dynamic functional connectivity for EEG-based person identification in diverse human states. IEEE Trans Inform Forensics Secur 14(12):3259–3272

9. Wilaiprasitporn T, Ditthapron A, Matchaparn K, Tongbuasirilai T, Banluesombatkul N, Chuangsuwanich E (2019) Affective EEG-based person identification using the deep learning approach. IEEE Trans Cogn Dev Syst 12(3):486–496
10. Ding Y, Robinson N, Tong C, Zeng Q, Guan C (2023) LGGNet: learning from local-global-graph representations for brain–computer interface. IEEE Trans Neural Netw Learn Syst 35(7)
11. Ying Z, You J, Morris C, Ren X, Hamilton W, Leskovec J (2018) Hierarchical graph representation learning with differentiable pooling. Adv Neural Inform Process Syst 31
12. Kipf TN, Welling M (2016) Semi-supervised classification with graph convolutional networks. arXiv preprint arXiv:1609.02907
13. Bai Y, Ding H, Bian S, Chen T, Sun Y, Wang W (2019) SimGNN: a neural network approach to fast graph similarity computation. In: Proceedings of the twelfth ACM international conference on web search and data mining, pp 384–392
14. Zheng W-L, Bao-Liang L (2015) Investigating critical frequency bands and channels for EEG-based emotion recognition with deep neural networks. IEEE Trans Autonom Mental Dev 7(3):162–175
15. Liu W, Qiu J-L, Zheng W-L, Lu B-L (2021) Comparing recognition performance and robustness of multimodal deep learning models for multimodal emotion recognition. IEEE Trans Cogn Dev Syst 14(2):715–729
16. Peng Y, Qin F, Kong W, Ge Y, Nie F, Cichocki A (2022) GFIL: a unified framework for the importance analysis of features, frequency bands, and channels in EEG-based emotion recognition. IEEE Trans Cog Dev Syst 14(3):935–947
17. Song T, Zheng W, Song P, Cui Z (2018) EEG emotion recognition using dynamical graph convolutional neural networks. IEEE Trans Affect Comput 11(3):532–541
18. Lawhern VJ, Solon AJ, Waytowich NR, Gordon SM, Hung CP, Lance BJ (2018) EEGNet: a compact convolutional neural network for EEG-based brain–computer interfaces. J Neural Eng 15(5):056013
19. Ding Y, Robinson N, Zeng Q, Chen D, Wai AAP, Lee T-S, Guan C (2020) TSception: a deep learning framework for emotion detection using EEG. In: 2020 international joint conference on neural networks (IJCNN). IEEE, pp 1–7

Open Access This chapter is licensed under the terms of the Creative Commons Attribution-NonCommercial-NoDerivatives 4.0 International License (http://creativecommons.org/licenses/by-nc-nd/4.0/), which permits any noncommercial use, sharing, distribution and reproduction in any medium or format, as long as you give appropriate credit to the original author(s) and the source, provide a link to the Creative Commons license and indicate if you modified the licensed material. You do not have permission under this license to share adapted material derived from this chapter or parts of it.

The images or other third party material in this chapter are included in the chapter's Creative Commons license, unless indicated otherwise in a credit line to the material. If material is not included in the chapter's Creative Commons license and your intended use is not permitted by statutory regulation or exceeds the permitted use, you will need to obtain permission directly from the copyright holder.

Chapter 4
Specific-Task and Multi-session Brain Fingerprint Identification with Joint Disentangled Representation

Abstract The rapid adoption of wearable devices has significantly expanded the potential applications of EEG signals in biometric identification, including brain fingerprint recognition. EEG, being non-invasive and offering real-time data acquisition, provides an attractive solution for secure authentication systems. However, a major obstacle in implementing EEG-based brain fingerprint recognition in real-world scenarios is the inherent variability present in cross-session EEG data. This variability, arising from factors such as changes in user state, environmental influences, and equipment variations, severely impacts the reliability and stability of models over time. In addition, the heterogeneity of task-specific protocols further complicates the generalization of models, making it difficult to ensure consistent performance across different sessions or individuals. Given the increasing demand for secure, long-term authentication systems, overcoming these challenges is essential for the practical deployment of EEG-based brain fingerprint recognition technologies. This innovative framework is specifically designed to enhance cross-session EEG-based brain fingerprint identification under varying task protocols. The JDR-DAT framework tackles the problem of variability by disentangling identity-related features using mutual information estimation, ensuring that the model can reliably extract unique user information irrespective of session-related changes. Furthermore, by incorporating domain adversarial training, the framework improves robustness across different sessions, enhancing the model's ability to maintain high accuracy over time. Extensive tests on longitudinal EEG data from two publicly accessible datasets, Rapid Serial Visual Presentation and Motor Imagery, validate the effectiveness of JDR-DAT, achieving average accuracies of 85.83 and 96.72%, respectively.

4.1 Introduction

EEG signals have been widely acknowledged for their suitability in biometric applications owing to their unique brainwave characteristics [1]. Recent progress in technology, especially with the rise of consumer-grade wearables, has significantly improved the practicality of hands-free authentication using EEG data obtained from dedicated sensors. This method offers a viable alternative in situations where conventional biometric techniques, such as fingerprint or facial recognition, may face limitations, particularly for individuals with physical impairments or in challenging environmental conditions [2]. Additionally, EEG-based biometric systems enhance privacy protection, as brainwave patterns are harder to replicate and can be easily reset by modifying the stimuli [3, 4].

The variability in EEG data across sessions presents considerable challenges for achieving stable and accurate brain fingerprint identification. Furthermore, the variety of single-task acquisition protocols, such as Visual Evoked Potentials (VEP) [5], Steady-State Visually Evoked Potentials (SSVEP) [6], Rapid Serial Visual Presentation (RSVP) [7], and Motor Imagery (MI) [8], enhances the dataset by capturing various facets of brain activity. However, this variety introduces difficulties in developing models that perform consistently and generalize well. Most existing methods tend to focus on a single protocol or session, which restricts their real-world applicability, where multiple protocols and sessions are common.

This chapter presents the Joint Disentangled Representation and Domain Adversarial Training (JDR-DAT) framework for EEG-based cross-session brain fingerprint identification using single-task protocols to tackle these issues. Our method starts by utilizing multi-scale convolutions and graph convolutional networks (GCNs) to extract complex features from raw EEG data. Instead of applying adversarial training to the full feature space, the extracted features are separated into three components: session-specific identity features, session-invariant identity features, and irrelevant nuisance variables. This disentanglement step separates identity-related EEG features that remain consistent across sessions from those affected by session-specific influences or variations in EEG acquisition protocols. Task labels corresponding to each protocol act as references, helping the model differentiate between identity-relevant information and irrelevant factors. This approach allows the model to handle protocol and session variations, improving its ability to learn session-invariant, identity-discriminative brainprint features. Additionally, domain adversarial training is used to align features across sessions, further boosting the model's robustness and generalizability.

This chapter extends prior work and presents two main contributions:

- A new framework, the Joint Disentangled Representation and Domain Adversarial Training (JDR-DAT), is introduced to extract improved session-invariant features with distinct identity-related characteristics.
- A quantitative analysis of EEG signals across five frequency bands within two different protocols is performed, offering valuable insights that may guide future studies in this area.

4.2 Joint Disentangled Representation with Domain Adversarial Training Framework

First, the problem of EEG-based cross-session brain fingerprint identification is defined, where each session is treated as a distinct domain. Let the labeled source domain be represented as $\tilde{\mathcal{D}}_S = \{X_S, Y_S\} = \{(x_S^1, y_S^1), \cdots, (x_S^{n_s}, y_S^{n_s})\}$, where n_s is the number of samples. Here, $x_S^i |_{i=1}^{n_s} \in \mathbb{R}^{d \times L}$ represents an EEG segment for a specific session, with d being the number of channels in the recording, and L denoting the sample length in time. The vector y_S^i belongs to $\mathbb{R}^C$, encoding the subject ID using one-hot encoding, where C is the number of subjects. The unlabeled target domain is denoted as $\tilde{\mathcal{D}}_T = \{X_T\} = \{x_T^1, \ldots, x_T^{n_t}\}$, where $x_T^i |_{i=1}^{n_t} \in \mathbb{R}^{d \times L}$ represents an EEG data segment for another session, and n_t is the number of samples in the target domain. Both $\tilde{\mathcal{D}}_S$ and $\tilde{\mathcal{D}}_T$ come from different sessions involving the same set of subjects but have different probability distributions. The goal is to design a deep learning model, denoted as $y = \theta(x)$, that minimizes the differences between the source and target domains while retaining the discriminative features necessary for effective brain fingerprint identification. To achieve this, the model aims to minimize the target risk $\epsilon_t(\theta) = Pr_{(x,y) \sim \tilde{\mathcal{D}}_T}[\theta(x) \neq y]$ using the labeled source domain data. An overview of the proposed framework is shown in Fig. 4.1.

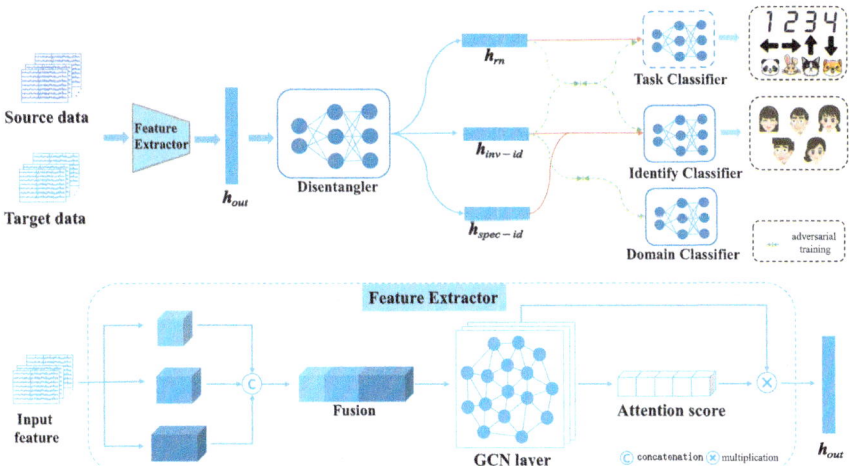

Fig. 4.1 The overview of the JDR-DAT framework is as follows: h_{out} denotes the embedding produced by the feature extraction module. This module utilizes adversarial training along with mutual information-based disentanglers to separate the embedding into three essential components: session-specific identity features ($h_{spec-id}$), session-invariant identity features (h_{inv-id}), and identity-irrelevant recording-related nuisance features (h_{rn}). At the same time, it minimizes the mutual information between these components to enhance the disentanglement process

4.2.1 Feature Extractor

This module extracts time-domain features from EEG signals using multi-scale convolutional kernels [9] and incorporates an attention mechanism to adjust the output of the subsequent graph neural network into embedding vectors.

Specifically, for a given set of EEG data, $x^i_{\text{input}} |_{i=1}^n \in \mathbb{R}^{d \times L} \sim \tilde{\mathcal{D}}_{S/T}$, multiple time convolutions are applied to emulate frequency filtering:

$$x^{(i,k)}_{T\_out} = \text{ReLU}\left(\mathcal{F}_{BN}\left(\mathcal{F}^k_{\text{Conv2D}}\left(x^i_{\text{input}}\right)\right)\right), \tag{4.1}$$

where $\mathcal{F}_{BN}(\cdot)$ denotes the batch normalization function, and $\mathcal{F}^k_{\text{Conv2D}}(\cdot)$ represents the 2D convolution operation. The kernel sizes are determined based on their proportional relationship to the sample duration L, represented by α_k for $k = 1, 2, 3$. Here, α_k takes the values $[0.1, 0.2, 0.5]$, indicating three different kernel types with corresponding sizes, denoted by A^k:

$$A^k = (1, \alpha_k \cdot L). \tag{4.2}$$

The output of the kth convolution is represented as $x^{(i,k)}_{T\_out} \in \mathbb{R}^{N \times d \times L_k}$, where N indicates the number of kernels, and L_k denotes the length of the feature map produced after convolution.

After applying the multi-scale convolutions, the resulting outputs are concatenated along the feature dimension to produce the final output of the multi-scale time convolutional layer, denoted as $x^i_{\text{cat}} \in \mathbb{R}^{N \times d \times \sum L_k}$:

$$x^i_{\text{cat}} = \Gamma(x^{(i,1)}_{T\_out}, \ldots, x^{(i,k)}_{T\_out}), \tag{4.3}$$

where the function $\Gamma(\cdot)$ represents the concatenation operation, combining the outputs from various scales along the feature dimension. Following this, a 1×1 convolution layer is applied to the concatenated features to generate the fused features $\bar{x}^i_{\text{fuse}} \in \mathbb{R}^{N \times d \times \sum L_k}$.

To capture the adjacency information between channels, $\bar{x}^i_{\text{fuse}}$ is flattened and transformed into a graph-based representation $x^i_{\text{fuse}} \in \mathbb{R}^{d \times N \cdot \sum L_k}$, where the nodes correspond to EEG channels. A Graph Convolutional Network (GCN), based on a general message passing framework [10], is then applied to x^i_{fuse}:

$$x_{\text{out}} = \text{GCN}(A, x^i_{\text{fuse}}), \tag{4.4}$$

where $A = x^i_{\text{fuse}} {x^i_{\text{fuse}}}^T \in \mathbb{R}^{d \times d}$ represents the adjacency matrix, which encodes the relationships between the channels.

Next, a global mean representation, $h_{\text{mean}} \in \mathbb{R}^{1 \times N \cdot \sum L_k}$, is computed by averaging the values in x_{out}. The similarity between each channel and h_{mean} is then calculated

4.2 Joint Disentangled Representation with Domain Adversarial Training Framework

to determine the significance of each channel. Finally, the weighted embeddings are combined to obtain the final embedding vector representation $\boldsymbol{h}_{\text{out}} \in \mathbb{R}^{1 \times N \cdot \sum L_k}$:

$$\boldsymbol{h}_{\text{out}} = \text{softmax}(\boldsymbol{h}_{\text{mean}} \boldsymbol{x}_{\text{out}}^T) \boldsymbol{x}_{\text{out}}. \tag{4.5}$$

4.2.2 Representation Disentanglement

In the proposed JDR-DAT, adversarial training is applied to achieve feature disentanglement. Initially, $\boldsymbol{h}_{\text{out}}$ is input into various disentanglers, collectively denoted as $\mathcal{F}_D(\cdot)$, each consisting of two fully connected layers. The goal of these disentanglers is to project $\boldsymbol{h}_{\text{out}}$ into separate feature spaces, isolating identity-related features $\boldsymbol{h}_{\text{id}} \in \mathbb{R}^{1 \times L_d}$ from recording-related nuisance information $\boldsymbol{h}_{\text{rn}} \in \mathbb{R}^{1 \times L_d}$. Here, L_d represents the length of the features after passing through the disentanglers, corresponding to the dimensionality of the disentangled feature space. Subsequently, the classifier $\mathcal{F}_{\text{C-id}}(\cdot)$ is trained to achieve accurate identification using cross-entropy loss:

$$\mathcal{L}_{c1} = -\mathbb{E}_{(x_s, y_s) \sim \tilde{\mathcal{D}}_s} \sum y_s \log(\mathcal{F}_{\text{C-id}}(\boldsymbol{h}_{\text{id}})). \tag{4.6}$$

To enhance the disentanglement performance, task labels y_{task} are incorporated to guide the model in isolating nuisance factors:

$$\mathcal{L}_{c2} = -\mathbb{E}_{x \sim \tilde{\mathcal{D}}_{s/t}} \sum y_{\text{task}} \log(\mathcal{F}_{\text{C-task}}(\boldsymbol{h}_{\text{rn}})). \tag{4.7}$$

With the classifiers fixed, the feature sets are employed to train the model in an adversarial manner, inducing confusion by maximizing the entropy of classifier predictions through negative entropy minimization:

$$\mathcal{L}_{\text{ent}} = \sum_{x \sim \tilde{\mathcal{D}}_{s/t}} \log(\mathcal{F}_{\text{C-id}}(\boldsymbol{h}_{\text{rn}})) + \sum_{x \sim \tilde{\mathcal{D}}_{s/t}} \log(\mathcal{F}_{\text{C-task}}(\boldsymbol{h}_{\text{id}})). \tag{4.8}$$

The first term aims to confuse the identity classifier $\mathcal{F}_{\text{C-id}}(\cdot)$ by using $\boldsymbol{h}_{\text{rn}}$, while the second term serves to confuse the task classifier $\mathcal{F}_{\text{C-task}}(\cdot)$ with $\boldsymbol{h}_{\text{id}}$.

For domain alignment, features across sessions are aligned using domain adversarial training:

$$\mathcal{L}_d = -\mathbb{E}_{x \sim \tilde{\mathcal{D}}_s}[\log(\mathcal{F}_{\text{C-domain}}(\boldsymbol{h}_{\text{id}}))] - \mathbb{E}_{x \sim \tilde{\mathcal{D}}_t}[\log(1 - \mathcal{F}_{\text{C-domain}}(\boldsymbol{h}_{\text{id}}))]. \tag{4.9}$$

The first term ensures accurate predictions for source domain data by the domain classifier $\mathcal{F}_{\text{C-domain}}$, while the second term encourages target domain features to deceive the domain classifier, thus fostering domain-invariant feature extraction.

Further disentanglement is achieved by splitting $\boldsymbol{h}_{\text{id}} \in \tilde{\mathcal{D}}_S$ into session-invariant identity features $\boldsymbol{h}_{\text{inv-id}} \in \mathbb{R}^{1 \times L_d}$ and session-specific identity features $\boldsymbol{h}_{\text{spec-id}} \in$

Algorithm 4.1 Learning algorithm for JDR-DAT

Input: a labeled source domain $\tilde{\mathcal{D}}_S$, an unlabeled target domain $\tilde{\mathcal{D}}_T$; feature extractor $\mathcal{F}_E$; disentangler $\mathcal{F}_D$; classifiers $\mathcal{F}_{\text{C-id}}$, $\mathcal{F}_{\text{C-task}}$, $\mathcal{F}_{\text{C-domain}}$.
Output: well-trained feature extractor $\tilde{\mathcal{F}}_E$, disentangler $\tilde{\mathcal{F}}_D$, classifiers $\tilde{\mathcal{F}}_{\text{C-id}}$.
1: **while** not converged **do**
2:    **Feature Extraction**:
3:    Extract features from $x \sim \tilde{\mathcal{D}}_{S/T}$ using multi-scale convolutional kernels to obtain $\boldsymbol{h}_{\text{out}}$ (Eqs. 4.1–4.5).
4:    **Disentanglement**:
5:    Feed $\boldsymbol{h}_{\text{out}}$ into the disentangler $\mathcal{F}_D$ to get $\boldsymbol{h}_{\text{id}}$ and $\boldsymbol{h}_{\text{rn}}$.
6:    Update $\mathcal{F}_E$, $\mathcal{F}_D$, $\mathcal{F}_{\text{C-id}}$ using the cross-entropy loss $\mathcal{L}_{c1}$ (Eq. 4.6).
7:    Update $\mathcal{F}_E$, $\mathcal{F}_D$, $\mathcal{F}_{\text{C-task}}$ using the cross-entropy loss $\mathcal{L}_{c2}$ to guide the separation of nuisance variables (Eq. 4.7).
8:    Adversarially train to induce confusion in $\mathcal{F}_{\text{C-id}}$ and $\mathcal{F}_{\text{C-task}}$ by minimizing negative entropy $\mathcal{L}_{\text{ent}}$ (Eq. 4.8).
9:    **Domain Adversarial Training**:
10:    Separate $\boldsymbol{h}_{\text{out}} \in \tilde{\mathcal{D}}_S$ into $\boldsymbol{h}_{\text{inv-id}}$ and $\boldsymbol{h}_{\text{spec-id}}$.
11:    Align features across sessions by updating $\mathcal{F}_E$, $\mathcal{F}_D$, and $\mathcal{F}_{\text{C-domain}}$ using domain adversarial training loss $\mathcal{L}_d$ (Eq. 4.9).
12:    Minimize prediction differences between $\boldsymbol{h}_{\text{inv-id}}$ and $\boldsymbol{h}_{\text{spec-id}}$ using loss $\mathcal{L}_{\text{is}}$ (Eq. 4.10).
13:    **Further Disentanglement**:
14:    Minimize mutual information between features using $\mathcal{L}_{\text{MI}}$ (Eq. 4.11).
15: **end while**
16: **return** $\tilde{\mathcal{F}}_E$, $\tilde{\mathcal{F}}_D$, $\tilde{\mathcal{F}}_{\text{C-id}}$.

$\mathbb{R}^{1 \times L_d}$. A supplementary loss term $\mathcal{L}_{\text{is}}$ is introduced to minimize the prediction difference between $\boldsymbol{h}_{\text{inv-id}}$ and $\boldsymbol{h}_{\text{spec-id}}$, both of which contain identity-related information:

$$\mathcal{L}_{\text{is}} = \|\mathcal{F}_{\text{C-id}}(\boldsymbol{h}_{\text{inv-id}}) - \mathcal{F}_{\text{C-id}}(\boldsymbol{h}_{\text{spec-id}})\|_1. \tag{4.10}$$

In addition, mutual information between features is minimized to further improve disentanglement:

$$\mathcal{L}_{\text{MI}}(x, z) = \mathcal{T}(x, z; \theta) - \log(e^{\mathcal{T}(x, z'; \theta)}), \tag{4.11}$$

where the function $\mathcal{T}(\cdot)$ is implemented as a fully connected network module with trainable parameters θ [11]. The pairs (x, z) are selected from the set $\{(\boldsymbol{h}_{\text{inv-id}}, \boldsymbol{h}_{\text{rn}}), (\boldsymbol{h}_{\text{inv-id}}, \boldsymbol{h}_{\text{spec-id}})\}$, and z' is sampled from the set $\{\boldsymbol{h}_{\text{rn}}, \boldsymbol{h}_{\text{spec-id}}\}$.

A more detailed explanation of the training process is provided in Algorithm 4.1. Additionally, Table 4.1 offers an in-depth, layer-by-layer breakdown of the network structure, outlining the input and output dimensions for each operation in the JDR-DAT framework.

Table 4.1 The primary network architecture of the JDR-DAT framework. This table describes the operations performed in each layer, along with the corresponding input and output sizes

Layer	Operations	Input size	Output size
Multi-scale convolution	Conv2D $N \times (1, \alpha_k \cdot L)$	(1, d, L)	(N, d, L_k)
	BatchNorm		
	ReLU		
1×1 convolution	Concatenation	(N, d, L_k)	(N, d, $\sum L_k$)
	Conv2D $N \times (1, 1)$	(N, d, $\sum L_k$)	(N, d, $\sum L_k$)
GCN	Reshape	(N, d, $\sum L_k$)	(d, N $\times \sum L_k$)
	Graph convolution	(d, N $\times \sum L_k$)	(d, N $\times \sum L_k$)
Disentanglers	Fully Connected (FC)	(1, N $\times \sum L_k$)	(1, L_d)
Identity classifier	Fully Connected (FC)	(1, L_d)	–
Task classifier	Fully Connected (FC)	(1, L_d)	–
Domain classifier	Fully Connected (FC)	(1, L_d)	–

4.3 Brain Fingerprint Identification with JDR-DAT

4.3.1 Data Pre-processing

This research employs two separate multi-session datasets, each designed for specific mental tasks. A summary of these datasets can be found in Table 4.2, with more detailed descriptions provided below:

- **RSVP-based dataset** (Dataset I): This dataset, detailed in Zheng et al. [12], consists of EEG recordings from 14 participants engaged in a target image detection task. Each participant took part in the experiment twice, with an average interval of around 23 d between sessions. The raw EEG data underwent bandpass filtering in the range of 1–50 Hz. Each input sample has a duration of 0.5 s.

Table 4.2 Description of datasets utilized in this paper, presenting diverse protocols, with each dataset characterized by specific numbers of channels, subjects, sample count per session, sample length, and sessions

Detail	Dataset I	Dataset II
Protocol	RSVP	MI
Channels	62	22
Subjects	14	9
Samples per session	9128	2592
Sample length (s)	0.5	3.0
Sessions	2	2

- **BCI Graz dataset A** (Dataset II): This dataset includes EEG recordings from 9 participants performing four distinct motor imagery tasks: imagining movements of the left hand, right hand, both feet, and the tongue. Each participant participated in two separate sessions. The raw EEG data was bandpass-filtered between 1 and 50 Hz, downsampled to 200 Hz, and each input sample lasts approximately 3 s, containing 688 data points.

It is important to note that for Dataset I, the sample length is 0.5 s, with the window starting 0.5 s after the onset of the target stimulus [12]. For Dataset II, each sample contains 688 data points post-downsampling, with the last 625 points corresponding to a duration of roughly 3 s.

4.3.2 Implementation Details

The proposed framework is implemented using the PyTorch library and utilizes the Adam optimizer with a learning rate of 1e-3. The convolutional kernel configurations are based on those in Ding et al. [9]. The network comprises 64 temporal convolutional layers, with the output embedding length of the graph network set to 512.

In addition to the proposed feature extraction network, we introduce three benchmark models commonly used in EEG-based BCI tasks for comparison: EEGNet [13], TSception [14], and Brainnet [15]. For all models, the training process involves 100 epochs, with a batch size of 64. In the context of cross-session brain fingerprint identification, this study adopts a longitudinal sequence such as "s1→s2", where session 1 is treated as the labeled source domain, and session 2 as the unlabeled target domain for prediction.

A significant portion of the experiments revolves around two key paradigms:

- **JDR-DAT Framework**: This approach leverages task labels as anchors, helping to improve the model's ability to effectively disentangle complex features.
- **Domain Adaptation**: Serving as our baseline, this paradigm demonstrates the use of an adversarial component in the feature subspace without incorporating a task classifier.

4.3.3 Performances of Specific-Task and Multi-session Brain Fingerprint Identification

The accuracy values and corresponding F1 scores for cross-session brain fingerprint identification are summarized in Tables 4.3 and 4.4. These results are based on the two paradigms described above and were obtained from datasets collected using different data acquisition protocols.

4.3 Brain Fingerprint Identification with JDR-DAT

Table 4.3 Recognition results (%) on **Dataset I** (RSVP-based dataset) and **Dataset II** (Motor Imagery dataset) in the **JDR-DAT Framework** paradigm

Model	Dataset I		Dataset II	
	ACC	F1	ACC	F1
EEGNet	71.37	71.18	69.33	67.82
BrainNet	72.80	70.46	77.26	73.59
TSception	83.72	83.78	90.55	90.54
Ours	**85.83**	**85.84**	**96.72**	**96.74**

Table 4.4 Recognition results (%) on **Dataset I** (RSVP-based dataset) and **Dataset II** (Motor Imagery dataset) in the **Domain Adaptation** paradigm

Model	Dataset I		Dataset II	
	ACC	F1	ACC	F1
EEGNet	58.59	58.17	72.02	67.33
BrainNet	66.68	65.01	75.19	71.76
TSception	64.73	63.42	83.10	82.73
Ours	**69.58**	**67.84**	**85.60**	**83.06**

It is noteworthy that across all models, the disentangling framework, which involves jointly training multiple adversarial units in the feature subspace, outperforms the use of a single adversarial unit. Specifically, for the RSVP dataset, the JDR-DAT framework with anchors improves the average accuracy by 13.53% compared to domain adaptation alone, achieving a peak accuracy of 85.83% and an F1 score of 85.84%. For the MI dataset, the average accuracy increases by 4.48%, with a peak accuracy of 96.72% and an F1 score of 96.74%. On average, the accuracy improvement across both datasets is 9.01%.

Furthermore, a Wilcoxon signed-rank test was performed to assess the performance of all algorithms across the two paradigms. The test yielded a p-value of 0.023, which is below the 0.05 significance level, indicating a statistically significant difference in both accuracy and F1 score between the two paradigms. These results underscore the substantial enhancement provided by our proposed method over standard domain adversarial approaches, in line with the findings reported in [16].

Confusion matrices are employed to assess and compare the classification performance for each subject across different experimental paradigms. Figure 4.2 presents the confusion matrices for the RSVP and MI datasets under two paradigms: the JDR-DAT Framework and Domain Adaptation. Figure 4.2a and c show the classification results using the JDR-DAT Framework, while Fig. 4.2b and d display the outcomes from Domain Adaptation.

From the analysis of these confusion matrices, it is evident that there are distinct differences in classification performance between the two paradigms across both

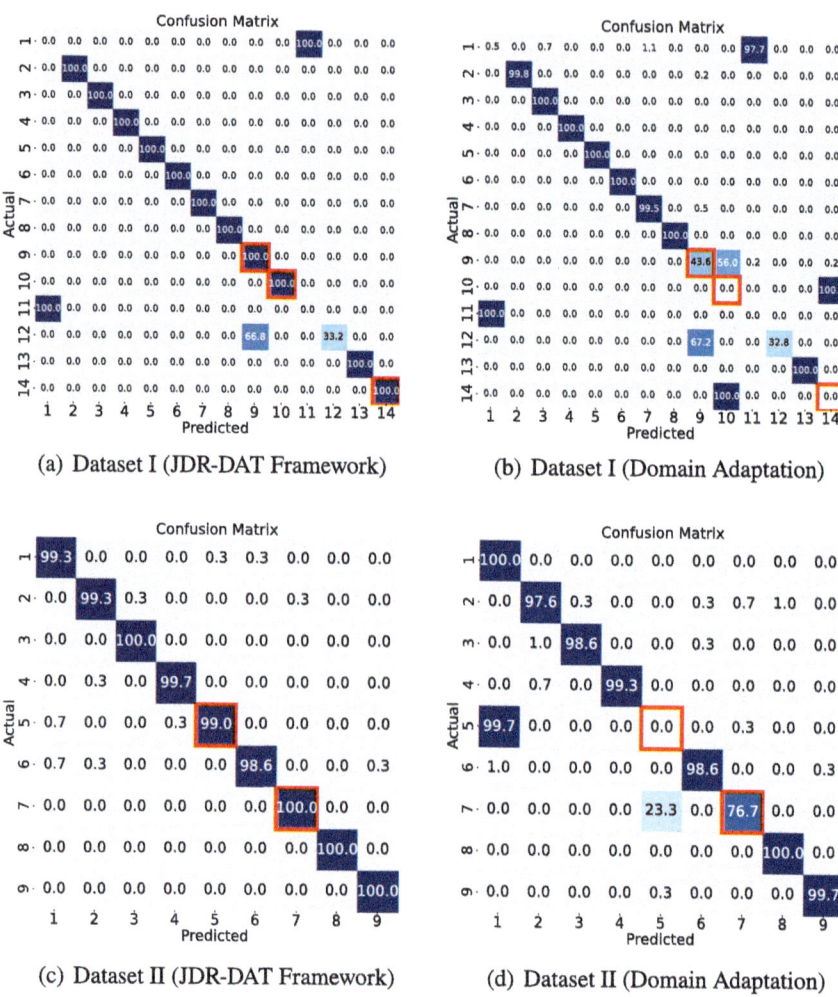

Fig. 4.2 Comparison of confusion matrices for Dataset I (RSVP) and Dataset II (MI) across two experimental paradigms: JDR-DAT framework and domain adaptation. The confusion matrices illustrate the classification performance under both paradigms, highlighting differences in model behavior across the datasets

datasets. In particular, the JDR-DAT Framework exhibits improved model robustness, leading to a reduction in classification errors when compared to Domain Adaptation.

4.3 Brain Fingerprint Identification with JDR-DAT

Table 4.5 Accuracy and F1 scores (%) comparison in the ablation study on RSVP and MI datasets

Setting	Description	RSVP		MI	
		ACC	F1	ACC	F1
Full model	Complete JDR-DAT framework	**85.83**	**85.84**	**96.72**	**96.74**
Model 1	No task labels as anchors	80.83	80.24	87.30	84.04
Model 2	No confusion introduced	63.53	62.73	86.95	83.82
Model 3	No minimization of prediction differences	76.21	72.42	65.93	61.48
Model 4	No mutual information estimation	64.26	63.48	88.46	84.94
Model 5	Domain adversarial baseline	69.58	67.84	85.60	83.06

4.3.4 Effect of Each Component

A series of ablation experiments are conducted to evaluate the robustness and efficiency of the proposed JDR-DAT framework. These experiments aim to assess the contributions of different loss components and modules integral to the overall framework. The ablation settings are as follows:

- **Full Model**: The complete JDR-DAT framework, incorporating all proposed components and loss functions.
- **Model 1**: This model excludes the loss component $\mathcal{L}_{c2}$ (Eq. 4.7), meaning task labels are not used as anchors to guide the disentanglement process.
- **Model 2**: This model omits the loss component $\mathcal{L}_{ent}$ (Eq. 4.8), removing the introduction of confusion to aid in disentangling identity-relevant features from task-specific variations.
- **Model 3**: This model excludes the norm loss component $\mathcal{L}_{is}$ (Eq. 4.10), which minimizes prediction differences between h_{inv-id} and $h_{spec-id}$.
- **Model 4**: This model omits the mutual information estimation loss $\mathcal{L}_{MI}$ (Eq. 4.11), which is essential for reducing redundancy between disentangled features.
- **Model 5**: This model serves as the domain adversarial baseline, excluding all disentangling components.

Table 4.5 presents the structural compositions of the models and summarizes the performance variations observed when different components of our proposed framework are modified across the two datasets. The removal of disentanglement-related components consistently leads to a decrease in overall model performance. In particular, components like $\mathcal{L}_{c2}$ and $\mathcal{L}_{ent}$ play a vital role in guiding the model and introducing necessary confusion for effective feature disentanglement. Omitting these components results in a significant drop in performance, sometimes even below the baseline. These experimental results underscore the importance of including all modules simultaneously to optimize model performance.

Table 4.6 Performance comparison (%) for different frequency bands on RSVP and MI datasets (Accuracy and F1 Scores)

Frequency band	RSVP		MI	
	ACC	F1	ACC	F1
Delta (1–4 Hz)	43.80	43.79	43.02	43.02
Theta (4–8 Hz)	44.96	45.40	81.75	79.66
Alpha (8–13 Hz)	69.38	67.38	80.40	80.40
Beta (13–30 Hz)	56.77	54.90	**86.19**	**85.81**
Gamma (30–50 Hz)	**83.58**	**83.52**	44.21	36.41

4.3.5 Effect of Critical Frequency Bands

To explore the contributions of different EEG signal frequency bands, the data are divided into five frequency bands using bandpass filters: Delta (1–4 Hz), Theta (4–8 Hz), Alpha (8–12 Hz), Beta (12–30 Hz), and Gamma (30–50 Hz). The experimental results on the two datasets are shown in Table 4.6.

The experimental results reveal substantial differences in EEG signal performance across various frequency bands for different tasks. Specifically, the Gamma band exhibited the highest performance in RSVP tasks, which aligns with previous research [17]. In contrast, the Theta, Alpha, and Beta bands performed better in MI tasks, consistent with studies on motor imagery [18, 19]. These findings emphasize the importance of selecting appropriate frequency bands for signal processing and feature extraction to improve the accuracy and reliability of EEG signal classification. Future research should explore combinations of different frequency bands and investigate finer subdivisions to identify optimal band selection strategies.

4.4 Conclusion

In this chapter, we have presented the Joint Disentangled Representation and Domain Adversarial Training (JDR-DAT) framework for EEG-based cross-session brain fingerprint identification within single-task protocols. By leveraging disentangled representations and domain adversarial training, the JDR-DAT framework effectively isolates session-invariant identity features while mitigating the influence of session-specific variations and identity-irrelevant noise.

The primary advantage of JDR-DAT lies in its ability to enhance the stability and reliability of EEG-based identification systems, particularly in real-world scenarios where session variability is a significant challenge. The framework's integration of mutual information estimation and adversarial training enables it to produce robust identity features that are less susceptible to the fluctuations caused by different sessions or protocols. This makes it a promising solution for long-term, scalable brain fingerprint identification systems.

However, while the framework demonstrates notable improvements over baseline domain adaptation methods, it is important to acknowledge its limitations. The reliance on single-task protocols restricts the framework's applicability in more complex, multi-task scenarios. Furthermore, while JDR-DAT shows impressive performance in handling session variability, its effectiveness could be influenced by the specific characteristics of the EEG datasets used, and further testing across diverse protocols is required to assess its generalizability.

Looking ahead, future work should explore the potential of combining multi-frequency band features for more comprehensive signal processing and classification. Additionally, the extension of JDR-DAT to multi-task and multi-session frameworks could offer further enhancements in performance and robustness, broadening its applicability to a wider range of EEG-based brain fingerprint identification tasks.

References

1. Stassen HH (1980) Computerized recognition of persons by EEG spectral patterns. Electroencephalogr Clin Neurophysiol 49(1–2):190–194
2. Maiorana E (2020) Deep learning for EEG-based biometric recognition. Neurocomputing 410:374–386
3. Wang M, El-Fiqi H, Hu J, Abbass HA (2019) Convolutional neural networks using dynamic functional connectivity for EEG-based person identification in diverse human states. IEEE Trans Inform Forensics Secur 14(12):3259–3272
4. Wang M, Wang S, Jiankun H (2022) Cancellable template design for privacy-preserving EEG biometric authentication systems. IEEE Trans Inform Forensics Secur 17:3350–3364
5. Zhao H, Chen Y, Pei W, Chen H, Wang Y (2021) Towards online applications of EEG biometrics using visual evoked potentials. Expert Syst Appl 177:114961
6. Debie E, Moustafa N, Vasilakos A (2021) Session invariant EEG signatures using elicitation protocol fusion and convolutional neural network. IEEE Trans Depend Sec Comput 19(4):2488–2500
7. Salimi N, Barlow M, Lakshika E (2020) Towards potential of n-back task as protocol and EEGNET for the EEG-based biometric. In: 2020 IEEE symposium series on computational intelligence (SSCI). IEEE, pp 1718–1724
8. Meng L, Jiang X, Huang J, Li W, Luo H, Wu D (2023) User identity protection in EEG-based brain-computer interfaces: supplementary material. IEEE Trans Neural Syst Rehabil Eng 31
9. Ding Y, Robinson N, Tong C, Zeng Q, Guan C (2023) LGGNet: learning from local-global-graph representations for brain-computer interface. IEEE Trans Neural Netw Learn Syst 1–14
10. Kipf TN, Welling M (2016) Semi-supervised classification with graph convolutional networks. arXiv preprint arXiv:1609.02907
11. Peng X, Huang Z, Sun X, Saenko K (2019) Domain agnostic learning with disentangled representations. In: International conference on machine learning, pp 5102–5112. PMLR
12. Zheng L, Sun S, Pei W, Gao X, Zhang L, Wang Y (2020) A cross-session dataset for collaborative brain-computer interfaces based on rapid serial visual presentation. Front Neurosci 14:579469
13. Lawhern VJ, Solon AJ, Waytowich NR, Gordon SM, Hung CP, Lance BJ (2018) EEGNet: a compact convolutional neural network for eeg-based brain–computer interfaces. J Neural Eng 15(5):056013

14. Ding Y, Robinson N, Zhang S, Zeng Q, Guan C (2022) TSception: capturing temporal dynamics and spatial asymmetry from EEG for emotion recognition. IEEE Trans Affect Comput 14(3)
15. Fallahi M, Strufe T, Arias-Cabarcos P (2023) BrainNet: improving brainwave-based biometric recognition with Siamese networks. In: 2023 IEEE international conference on pervasive computing and communications (PerCom). IEEE, pp 53–60
16. Han M, Özdenizci O, Wang Y, Koike-Akino T, Erdoğmuş D (2020) Disentangled adversarial autoencoder for subject-invariant physiological feature extraction. IEEE Signal Process Lett 27:1565–1569
17. Herrmann CS, Munk MHJ, Engel AK (2004) Cognitive functions of gamma-band activity: memory match and utilization. Trends Cogn Sci 8(8):347–355
18. Al-Saegh A, Dawwd SA, Abdul-Jabbar JM (2021) Deep learning for motor imagery EEG-based classification: a review. Biomed Signal Process Control 63:102172
19. Pfurtscheller G, Brunner C, Schlögl A, Lopes Da Silva FH (2006) Mu rhythm (de) synchronization and EEG single-trial classification of different motor imagery tasks. NeuroImage 31(1):153–159

Open Access This chapter is licensed under the terms of the Creative Commons Attribution-NonCommercial-NoDerivatives 4.0 International License (http://creativecommons.org/licenses/by-nc-nd/4.0/), which permits any noncommercial use, sharing, distribution and reproduction in any medium or format, as long as you give appropriate credit to the original author(s) and the source, provide a link to the Creative Commons license and indicate if you modified the licensed material. You do not have permission under this license to share adapted material derived from this chapter or parts of it.

The images or other third party material in this chapter are included in the chapter's Creative Commons license, unless indicated otherwise in a credit line to the material. If material is not included in the chapter's Creative Commons license and your intended use is not permitted by statutory regulation or exceeds the permitted use, you will need to obtain permission directly from the copyright holder.

Chapter 5
Multi-task and Single-Session Brain Fingerprint Identification with Brain Network

Abstract In recent years, brain fingerprint identification has emerged as a promising biometrics modality due to its potential for providing high security and precision in identity recognition. Leveraging the unique neural patterns that emerge during cognitive tasks, this method holds particular promise for applications in high-security environments, such as defense, healthcare, and finance. In this chapter, an advanced brain network-based approach is proposed for multi-task and single-session brain fingerprint identification. The methodology begins with the construction of brain functional networks by calculating phase synchronization values between electroencephalography (EEG) channels, which allows for the capture of intricate brain activity patterns. Following this, a range of network metrics, such as node degree, clustering coefficient, and global efficiency, are computed to generate a comprehensive and multidimensional feature vector, effectively representing the complex dynamics of brain function. To ensure robust classification, Linear Discriminant Analysis (LDA) is employed to process these extracted features, enabling precise brain fingerprint identification with high accuracy. The proposed method is rigorously evaluated across four diverse datasets that encompass a wide array of cognitive tasks, providing a comprehensive testbed for evaluating the method's effectiveness. Experimental results demonstrate that the method achieves an average identification accuracy exceeding 95% across all datasets, with a peak accuracy of 99%. These findings highlight the method's robust potential for real-world brain fingerprint identification applications, illustrating its efficacy in providing accurate and reliable identity recognition even in the context of diverse, multi-task environments.

5.1 Introduction

Previous[1] studies have established that EEG signals serve as a powerful tool for brain fingerprint identification. However, these studies predominantly rely on amplitude information, which is inherently problematic due to its susceptibility to random fluctuations. Such fluctuations often introduce deviations in experimental outcomes, undermining the reliability of amplitude-based methods. A strong relationship between cognitive brain activity and neuronal synchronization has been observed [1]. Notably, if synchronization metrics do not correlate with amplitude, methods based solely on amplitude fail to provide a comprehensive analysis of EEG signals. Phase synchronization analysis offers a robust alternative by examining the interrelations between EEG channel pairs through phase and synchronization angles. Unlike amplitude-based approaches, phase synchronization focuses on the instantaneous phase of signals, effectively minimizing the influence of amplitude variations. This method provides a stable and reliable measure for studying EEG data. Research has demonstrated that phase synchronization, particularly in chaotic systems, is a valuable tool in neural sciences. Additionally, it serves as an effective technique for assessing functional connectivity in EEG signals, making it broadly adopted in biological and neuroscientific research. Functional networks are hypothesized to emerge from coherent electrophysiological activities spanning multiple brain regions [2]. EEG phase synchronization-based brain networks have been extensively utilized in disease monitoring and prediction. For instance, Ling et al. [3] employed phase synchrony index matrix to construct brain networks, revealing a pronounced loss of small-world characteristics in patients compared to healthy individuals. Similarly, Jamal et al. [4] derived synchrostate features from EEG signals to quantify phase synchronization and build brain connectivity graphs. Despite these advances, task-specific datasets remain the predominant approach for hypothesis validation. This limitation arises because existing methods are often tailored to handle only specific types of EEG data effectively. In such experiments, participants are typically required to perform a singular type of task or remain in controlled environments to ensure data consistency [5].

This chapter presents an EEG-based brain fingerprint identification method that leverages phase synchronization analysis. The identification process utilizes brain functional networks derived from phase synchronization values. Specifically, the connectivity matrix, generated from the Phase Locking Value (PLV) matrix, is employed to construct a weighted, undirected network. Key network attributes, such as node degree, clustering coefficient, and global efficiency, are aggregated to characterize individuals [6]. These attributes effectively capture distinct patterns of brain activity, making them highly suitable for brain fingerprint identification. Unlike existing studies that predominantly rely on amplitude information, our approach focuses on

---

[1] This chapter incorporates content from "Personal Identification Based on Brain Networks of EEG Signals" by Wanzeng Kong, et al., available under CC BY-NC-ND 4.0. The original work is accessible at https://dl.acm.org/doi/abs/10.2478/amcs-2018-0057. The content has been slightly modified for integration.

5.2 Brain Network

phase synchronization of EEG signals. Phase synchronization describes the potential interrelationships between pairs of EEG channels, offering a robust alternative to amplitude-based methods. Furthermore, the proposed method is evaluated using four distinct datasets involving various tasks, thereby extending beyond the conventional approach of analyzing a single type of EEG signal from a specific task. This comprehensive evaluation highlights the generalizability and effectiveness of the method across diverse experimental conditions.

5.2 Brain Network

This section provides a detailed explanation of brain fingerprint identification using phase synchronization analysis. An overview of the proposed framework is depicted in Fig. 5.1.

5.2.1 Pre-processing

EEG signals are inherently non-stationary and are highly susceptible to external noise, which can significantly impact signal quality. Pre-processing the raw EEG data is essential to mitigate these disturbances. Initially, EEG data are filtered within

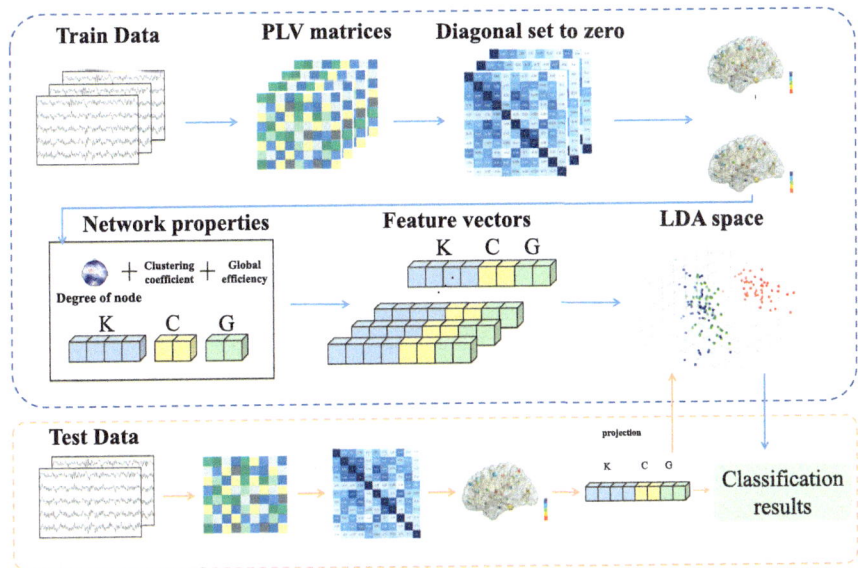

Fig. 5.1 Specific process of the proposed method

the frequency range of 2–47 Hz to remove irrelevant components. Each channel's values are then processed using the Common Average Reference (CAR) method [7], which enhances signal quality by reducing spatial noise. Subsequently, Butterworth bandpass filters are applied to isolate specific frequency bands of interest, ensuring the retention of meaningful neural oscillations.

5.2.2 Phase Synchronization

Phase synchronization provides a robust approach for analyzing EEG signals by emphasizing the phase relationships between channels while effectively minimizing the influence of amplitude variations. Synchronization is quantified using the Phase Locking Value (PLV), a widely recognized metric in neural signal processing.

The PLV for two continuous time series, $x(t)$ and $y(t)$, is defined as:

$$PLV = |<\exp(j\{\Phi_x(t) - \Phi_y(t)\})>|, \tag{5.1}$$

where $<.>$ denotes the averaging operator over continuous time t. The terms $\Phi_x(t)$ and $\Phi_y(t)$ represent the instantaneous phases of $x(t)$ and $y(t)$, respectively, at a given time t.

There are two primary methods for calculating the instantaneous phase: the Hilbert transform and the Gabor wavelet transform [8]. According to Mcfarland et al. [7], these two methods yield comparable results in processing EEG signals. In this study, the Hilbert transform is employed to compute the instantaneous phase due to its mathematical simplicity and computational efficiency. The Hilbert transform of a continuous time series $x(t)$ is defined as:

$$\tilde{x}(t) = \frac{1}{\pi} P \int_{-\infty}^{+\infty} \frac{x(t)}{t - \tau} d\tau, \tag{5.2}$$

where P denotes the Cauchy principal value, ensuring the integral avoids singularities when $t = \tau$. The signal $x(t)$ is assumed to be a narrowband signal, with its spectrum confined to a limited frequency range. This allows the envelope and phase to evolve gradually, enabling the extraction of meaningful phase information. The analytic signal, represented as $\tilde{x}(t)$, is derived from $x(t)$.

Using the Hilbert transform, $x(t)$ can be expressed as an analytic signal:

$$Z_x(t) = x(t) + j\tilde{x}(t) = A_x(t)e^{j\Phi_x(t)}, \tag{5.3}$$

where $A_x(t)$ and $\Phi_x(t)$ denote the instantaneous amplitude and phase of $x(t)$ respectively. The instantaneous phase $\Phi_x(t)$ is calculated as:

5.2 Brain Network

$$\Phi_x(t) = \arctan \frac{\tilde{x}(t)}{x(t)}. \tag{5.4}$$

Similarly, $\Phi_y(t)$ represents the instantaneous phase of another signal $y(t)$.

To quantify phase synchronization between $x(t)$ and $y(t)$, the phase difference $\Phi_{xy}(t)$ is defined as:

$$\Phi_{xy}(t) = |n\Phi_x(t) - m\Phi_y(t)| \leqslant \text{const}, \tag{5.5}$$

where n, m are integers, and const is a constant threshold for the phase difference. When this condition is satisfied, $x(t)$ and $y(t)$ are said to exhibit $n : m$ phase synchronization. In the experiments described in this chapter, const is set to 0.035, following the reference from [9]. The 1 : 1 phase synchronization is particularly relevant for neurobiological signals, as demonstrated in Kong's research on EEG [10]. Consequently, 1 : 1 phase synchronization is employed in this chapter to ensure consistency with established practices in EEG signal analysis.

Based on the instantaneous phase value at time t, the PLV between two-time series $x(t)$ and $y(t)$ can be computed [11]. A time window of 1 s is used for PLV calculation across different frequency bands, ensuring temporal resolution while maintaining computational efficiency. Each sample is divided into H non-overlapping time segments, and the average PLV is determined by averaging the phase-locking values across these H seconds. The average PLV is mathematically defined as:

$$PLV_{avg} = \frac{1}{H} \left| \sum_{h=1}^{H} \langle \exp(j\Delta\Phi) \rangle \right|, \tag{5.6}$$

where $\Delta\Phi$ represents the phase difference between $\Phi_x(t)$ and $\Phi_y(t)$ i.e., $\Delta\Phi = \Phi_x(t) - \Phi_y(t)$. The index $h = 1, 2, ..., H$ denotes the segment number, and $\langle . \rangle$ signifies the time averaging operation within each segment.

In this analysis, every channel is considered as an individual time series, and the phase synchronization is computed for every pair of N EEG channels. This process results in an $N \times N$ symmetric matrix V (Eq. 5.7), where each element v_{ij} corresponds to the average PLV between the ith and jth channels.

$$\mathbf{V} = \begin{bmatrix} 1 & v_{12} & \cdots & v_{1n} \\ v_{21} & 1 & \cdots & v_{2n} \\ \vdots & \vdots & \vdots & \vdots \\ v_{(n-1)1} & v_{(n-1)2} & \cdots & v_{(n-1)n} \\ v_{n1} & v_{n2} & \cdots & 1 \end{bmatrix}, \tag{5.7}$$

The symmetric matrix $\mathbf{V}$ serves as the connectivity matrix of the brain network, providing a quantitative representation of functional interactions between EEG channels. As shown in Fig. 5.2.

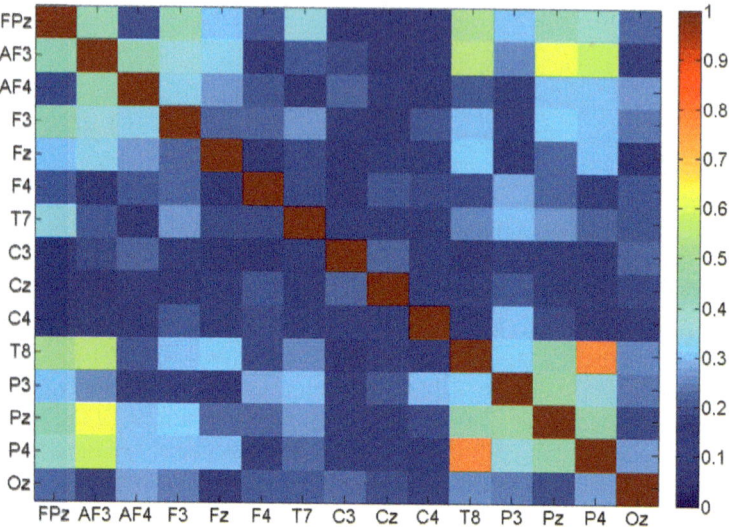

Fig. 5.2 The symmetric PLV matrix V consist of 15 channels. The color scale ranges from 0 to 1, where a chroma closer to 1 indicates greater synchronization between the EEG signals of two channels

5.2.3 Construction of Brain Network

A brain network characterizes the interactions among neurons, neuronal clusters, and brain regions. Based on connectivity types, brain networks can be classified into three categories: anatomical, functional, and effective connectivity [12, 13]. Functional connectivity, which captures statistical dependencies between spatially distributed neuronal activity, is a key focus in EEG analysis. Methods for evaluating functional connectivity are broadly divided into linear and nonlinear approaches [14].

In the context of this study, the brain network is modeled as a graph, leveraging graph theory for its analysis [15]. The graph is derived directly from the connectivity matrix obtained through phase synchronization analysis. Specifically, the Phase Locking Value (PLV) is employed to quantify functional connectivity, following a methodology similar to that described in Chavez et al. [16].

To eliminate self-connections, the diagonal elements of the connectivity matrix are set to zero. Using this adjusted connectivity matrix, a weighted undirected graph G is constructed. Formally, G is defined as $G = \{N, E, W\}$, where N represents the set of nodes, corresponding to EEG channels. E denotes the set of undirected edges, representing connections between nodes. W contains the weights that quantify the connection strengths. To ensure that the full spectrum of phase synchronization information is retained, no threshold is applied to the connectivity matrix, resulting in a fully connected weighted undirected graph G.

5.2.3.1 Network Analysis

Upon computing the PLV, a $N \times N$ weighted undirected graph is constructed using the connectivity matrix, where N represents the number of EEG channels. In this graph, each node corresponds to an EEG channel, and edges signify phase synchronization between channel pairs. The edge weights represent the synchronization strength. These graph attributes play a crucial role in analyzing the brain network using graph-theoretical measures.

- **Degree of Node**
  The degree is a fundamental metric in graph theory, describing the total connectivity of a node. In a weighted undirected graph G, the degree of a node quantifies the sum of weights ($w_{ij} \in W$) of its edges to other nodes. Specifically, the degree of node i is defined as:
  $$k_i^w = \sum_{i,j \in N, i \neq j} w_{ij}, \tag{5.8}$$
  where w_{ij} represents the weight of the edge connecting nodes i and j.

- **Global Efficiency**
  The global efficiency evaluates the efficiency of information transfer across the entire network [17]. Higher global efficiency indicates reduced costs for information exchange, signifying better network traffic capacity [18]. It is computed as the reciprocal of the harmonic mean of all shortest paths in the network [19]:
  $$E = \frac{1}{n-1} \sum_{i \in N} \frac{\sum_{j \in N, j \neq i} (d_{ij}^w)^{-1}}{n-1}, \tag{5.9}$$
  where d_{ij}^w is the shortest weighted path between nodes i and j. In the weighted network, $d_{ij}^w = \sum_{a \in g_{i \leftrightarrow j}^w} f(a)$, with $f(a)$ representing a mapping (e.g., inverse of weight) and $g_{i \leftrightarrow j}^w$ denoting the shortest path.

- **Clustering Coefficient**
  The clustering coefficient measures the tendency of a node's neighbors to form tightly-knit clusters [20]. As a key parameter, it reflects local connectivity within the network. The clustering coefficient for the entire graph is given as:
  $$C = \frac{1}{n} \sum_{i \in N} \frac{2t_i}{k_i(k_i - 1)}, \tag{5.10}$$
  where t_i denotes the geometric mean of the triangle weights around node i. Specifically, $t_i = \frac{1}{2} \sum_{j,h \in v_i} (w_{ij} w_{jh} w_{ih})^{1/3}$, with v_i representing the set of neighbors of node i, and k_i being the degree of node i [21].

The feature vector of each sample comprises the degree of nodes, global efficiency, and clustering coefficient. These graph-theoretical metrics can be efficiently computed using the Brain Connectivity Toolbox (http://www.brainconnectivity-toolbox.net/) [6].

5.2.4 Linear Discriminant Analysis

LDA, also known as Fisher Discriminant Analysis, is a widely used supervised classification technique based on linear discriminant functions [22]. The core idea of LDA is to project high-dimensional data into an optimal subspace that maximizes class separability [23]. It achieves this by maximizing the distance between classes while minimizing the variance within each class [24].

Given a dataset with M samples divided into N classes, each sample is represented by a feature vector of dimension d. Each feature vector is defined as $x_i = \{x_1^i, x_2^i, x_3^i, ..., x_d^i\}, i = \{1, 2, ..., N\}$. These samples are grouped into N classes, and a projection function $y = w^T x$ is used to map the samples into a subspace. The objective is to find the optimal w that maximizes inter-class separability. For multi-class problems, a k-dimensional projection matrix W is employed, and the projected data is given by $Y = W^T x$.

The mean of each class is defined as:

$$m_i = \frac{1}{M_i} \sum_{x \in N_i} x, \quad (5.11)$$

where N_i denotes class i, and M_i represents the sample numbers in N_i.

The scatter matrix for class i is defined as:

$$S_i = \sum_{x \in N_i} (x - m_i)(x - m_i)^T. \quad (5.12)$$

The within-class (S_w) and between-class (S_B) scatter matrices are then defined as:

$$S_w = \sum_{i=1}^{N} S_i = \sum_{i=1}^{N} \sum_{x \in N_i} (x - m_i)(x - m_i)^T, \quad (5.13)$$

$$S_B = \sum_{i=1}^{N} M_i (m_i - m)(m_i - m)^T, \quad (5.14)$$

where m is the global mean vector, calculated as the mean of all samples.

5.2 Brain Network

The discriminant criterion $J(W)$ is defined as:

$$J(W) = \frac{W^T S_B W}{W^T S_w W}. \tag{5.15}$$

The objective is to find W that maximizes $J(W)$, expressed as:

$$\begin{cases} W = \text{argmax}_w J(W) = \text{argmax}_w \frac{W^T S_B W}{W^T S_w W} \\ S.t. \quad W^T S_w W = c \neq 0 \end{cases} \tag{5.16}$$

To solve the optimization problem, a Lagrange multiplier is introduced, leading to:

$$L(w_i, \lambda) = w_i^T S_B w_i - \lambda(w_i^T S_w w_i - c). \tag{5.17}$$

The condition of getting the extremum of Eq. (5.16) is that the derivation of w_i is zero. Thus,

$$S_B w_i = \lambda S_w w_i, \tag{5.18}$$

$$S_w^{-1} S_B w_i = \lambda w_i, \tag{5.19}$$

The matrix W is composed of the feature vectors associated with k generalized feature values of $S_w^{-1} S_B$, where $k \leq N - 1$. Thus, $W = \{w_1, w_2, ..., w_k\}$.

Once the EEG data is projected into the lower-dimensional subspace, classification is performed using the nearest neighbor rule.

The proposed method utilizes average Phase Locking Value (PLV) to weight the brain's functional connectivity network and combines network attributes into feature vectors for individual identification. The main steps are as follows:

- **Step 1. Data Acquisition and Pre-processing**: Raw EEG data are collected from experimental datasets, followed by preprocessing and filtering into four frequency bands.
- **Step 2. PLV Computation**: The PLV of the preprocessed EEG signals is computed for each one-second non-overlapping time segment, resulting in H samples. The average PLV is calculated for each segment, forming an $N \times N$ symmetric matrix V.
- **Step 3. Functional Connectivity Network Construction**: The symmetric matrix V is used as the connectivity matrix to represent the brain's functional network. To eliminate self-connections, the diagonal elements of V are set to zero, producing a modified matrix V'. This matrix is then used to construct a weighted undirected graph G that models the brain's functional connectivity.
- **Step 4. Feature Extraction**: Key network attributes, including node degree, clustering coefficient, and global efficiency, are calculated for each sample. These attributes are combined into a feature vector for subsequent classification.

- **Step 5. Classification Using LDA**: The LDA projection space is learned from the training data. Test data are projected into the LDA space, and the nearest neighbor rule is applied to assign the class labels to test samples.

5.3 Brain Fingerprint Identification with Brain Network

This section presents an analysis of brain fingerprint identification using various datasets, including BCI Graz Dataset A, Motor Imagery Dataset, Neuromarketing Dataset (NMK), and Fatigue Driving Dataset (DRI). The differences in classification accuracy across the beta, gamma, alpha, and theta frequency bands are investigated. For each dataset, 480 s of EEG recordings per subject were divided into non-overlapping 30-s segments, resulting in 16 samples per subject. The mean PLV matrix for each segment was computed, and brain network attributes, such as node degree, global efficiency, and clustering coefficient, were extracted and combined into feature vectors. These samples were evenly split into training and testing sets, and classification was performed using the LDA classifier.

BCI Graz dataset A and Motor Imagery Dataset are publicly available datasets, hereafter referred to as the BCI data and Motor Imagery Dataset, respectively. The remaining two were recorded by our laboratory. The one obtained through a neuromarketing experiment would be referred to as the NMK data. The other one obtained from a complex task during a fatigue driving experiment is the DRI data.

5.3.1 Attributes Analysis

To evaluate the effectiveness of brain network attributes in brain fingerprint identification, this section analyzes the performance of three key attributes: node degree, global efficiency, and clustering coefficient. Statistical methods and topographic maps are used to illustrate these attributes.

5.3.1.1 Degree of Node

In the constructed brain network, each node corresponds to an EEG channel. The degree of each node reflects the strength of connectivity within the brain network. For visualization, topographic maps representing the average node degrees across subjects are generated using EEG signals.

For the Motor Imagery Dataset, as shown in Fig. 5.3, the electrode placement does not follow the standard 10–20 system, except for positions C3, Cz, and C4. Instead, the electrodes are positioned equidistantly, with a spacing of 2.5 cm between neighboring electrodes. Due to this unique layout, brain topographic maps for the

5.3 Brain Fingerprint Identification with Brain Network

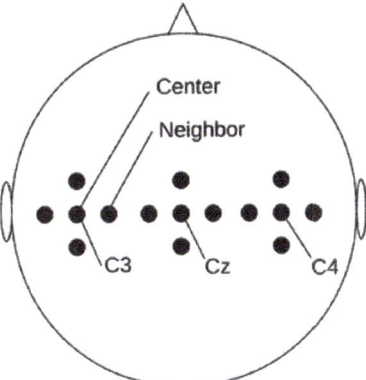

Fig. 5.3 Small Laplacian electrode placing scheme centered at C3, Cz, and C4. Distances between neighboring electrodes are 2.5 cm

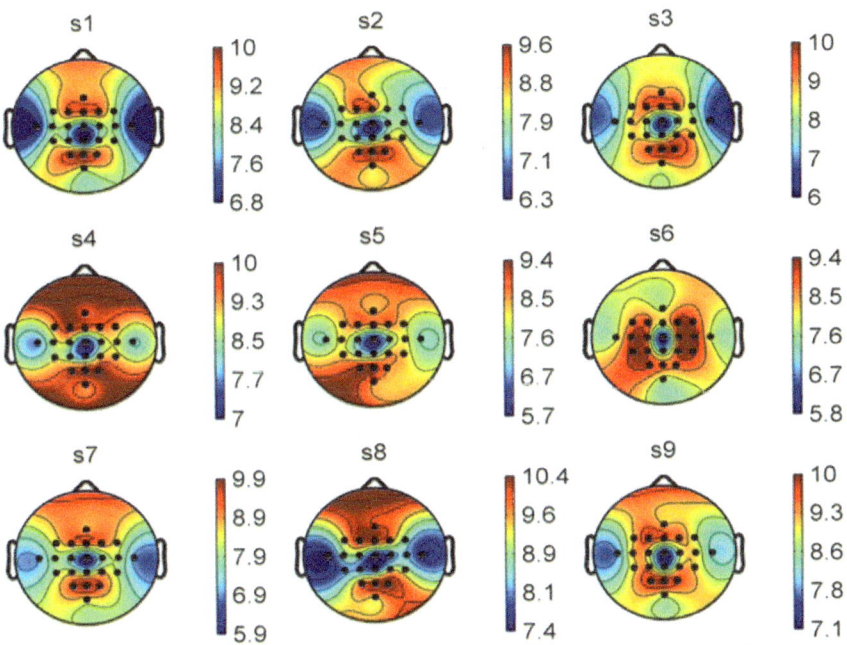

Fig. 5.4 Brain topographic map of four-class motor imagery tasks in BCI competition 2008 (BCI dataset). The black dots indicate the locations of the EEG channels, and the color represents the mean node degree for each EEG channel

Motor Imagery Dataset cannot be directly generated. Figures 5.4, 5.5 and 5.6 illustrate the topographic maps for the BCI, NMK, and DRI datasets.

From Figs. 5.4, 5.5 and 5.6, the brain topographic maps for the BCI, NMK, and DRI datasets reveal a certain degree of symmetry in connection strength between

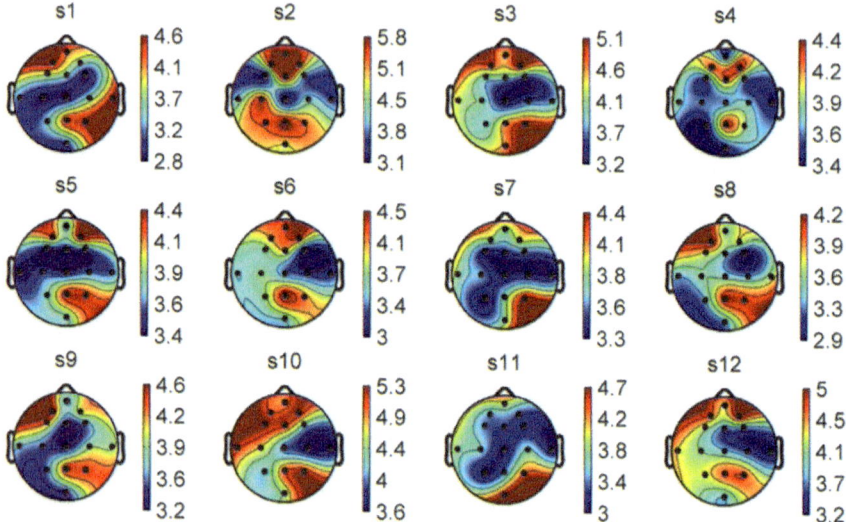

Fig. 5.5 The brain topographic map for the fatigue driving dataset (DRI dataset) recorded in our laboratory. The black dots indicate the locations of the EEG channels, with the color representing the mean node degree for each EEG channel

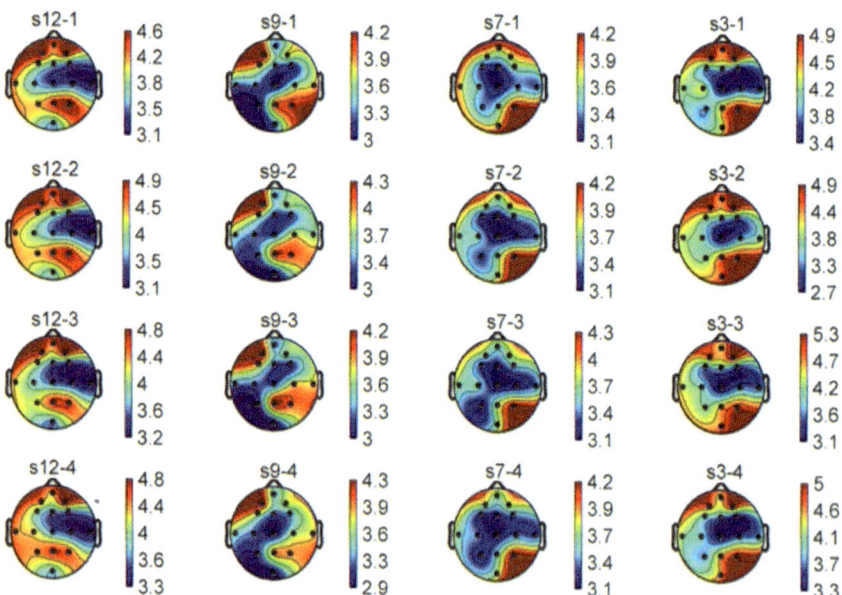

Fig. 5.6 Brain topographic map for a neuromarketing cognitive task recorded in our laboratory (NMK dataset). The black dots represent the locations of the EEG channels, with the color indicating the mean node degree for each EEG channel

5.3 Brain Fingerprint Identification with Brain Network

the left and right hemispheres for individual subjects. However, notable differences in the distribution of connection strength are observed across different subjects. In particular, the brain topographic maps for the DRI dataset, displayed in Fig. 5.5, highlight variations in connectivity patterns. The fatigue driving experiment involves complex tasks, and individuals with different knowledge levels or habitual responses may exhibit distinct cognitive processes when faced with the same challenge. This leads to differences in the topographic maps of some subjects, while others (e.g., subject 1 and subject 2) show similar patterns.

To enhance the reliability of the analysis, a statistical examination of node degrees is conducted. Using the BCI dataset as an example, the mean node degree vector for each of the 9 subjects is calculated, resulting in one mean vector per subject. A variance matrix is then computed across these 9 subjects. The results are presented in Fig. 5.7.

As depicted in Fig. 5.7, significant variations in node degrees are observed among individuals, emphasizing individual differences in brain network connectivity.

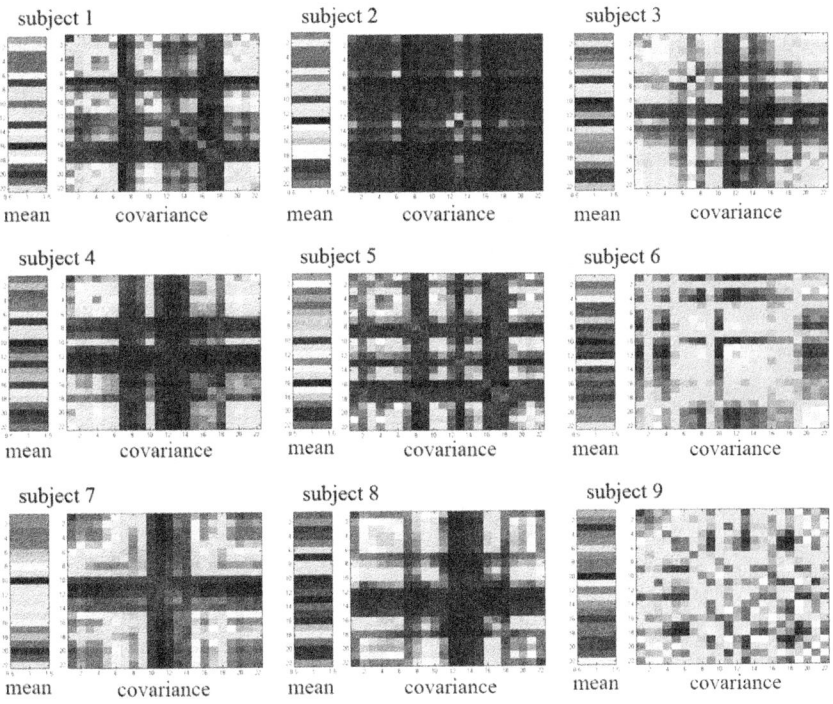

Fig. 5.7 The mean vector and covariance matrix for the four-class motor imagery tasks in BCI competition 2008 (BCI dataset)

5.3.1.2 Global Efficiency and Clustering Coefficient

Global efficiency and clustering coefficient are critical features in the construction of the feature vector. To further evaluate their impact and performance, the relationship between global efficiency and clustering coefficient for each subject is visualized in a scatter plot, as shown in Fig. 5.8. In the plot, global efficiency is represented on the x-axis, and clustering coefficient is represented on the y-axis.

Figure 5.8 illustrates the relationship between global efficiency and clustering coefficient across all datasets. Each subfigure corresponds to a dataset, and within each subfigure, individual subjects demonstrate distinct distribution patterns. A linear fit is applied to the data for each subject, revealing variation in the slopes and intercepts of the fitted lines. These results indicate that global efficiency and clustering coefficient can serve as unique identifiers for individuals. Moreover, the linear fits suggest a near-proportional relationship between global efficiency and clustering coefficient for most samples.

The observed relationship aligns with theoretical findings in prior studies. As noted in Latora and Marchiori [19], the clustering coefficient approximates efficiency on a local scale, while local efficiency is defined as the global efficiency calculated within the neighborhood of a given node [6]. Additionally, prior work

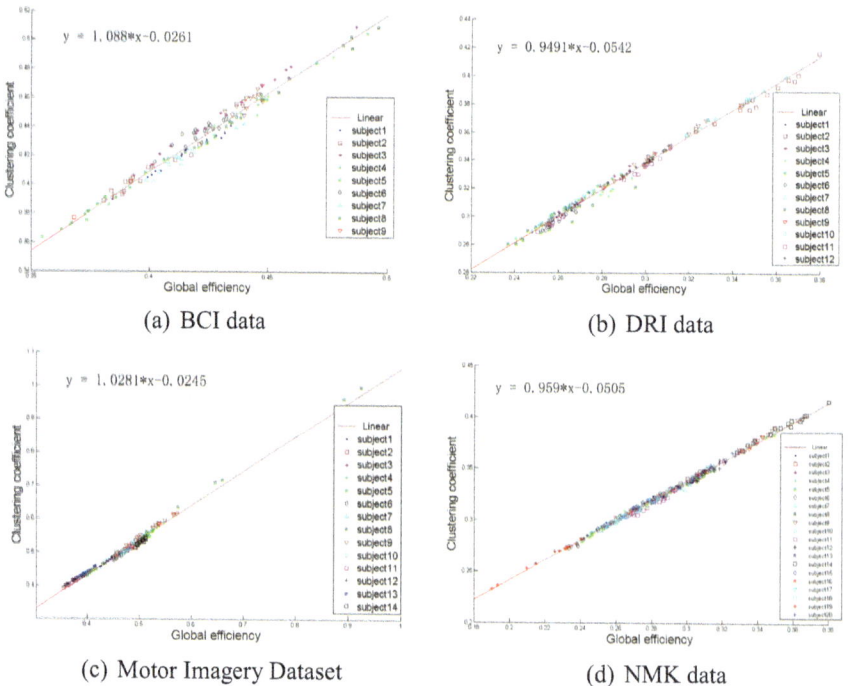

Fig. 5.8 Relationship between global efficiency and clustering coefficient

has shown that the weighted local efficiency closely corresponds to the weighted clustering coefficient [21]. These theoretical insights suggest that global efficiency and clustering coefficient exhibit similar patterns of variation, which is consistent with the experimental results obtained in this study.

In summary, the near-proportional relationship and the consistent variation patterns observed in the current experiment provide further evidence for the utility of global efficiency and clustering coefficient as discriminative features for brain fingerprint identification.

5.3.2 Performances of Multi-task and Single-Session Brain Fingerprint Identification

To evaluate the proposed method, a twofold cross-validation scheme is adopted for each dataset. The samples are randomly split into training and testing sets, and ten trials are conducted to calculate the average classification accuracy. Table 5.1 presents the classification accuracies in the beta band for the four datasets.

As observed in Table 5.1, the proposed method achieves consistently high classification accuracies across datasets. Specifically, accuracies for the BCI, DRI, and Motor Imagery datasets exceed 0.99 in all ten trials, while the NMK dataset achieves an accuracy exceeding 0.95. The minimal variations across trials further highlight the robustness of the proposed features for personal identification. These results confirm the effectiveness of brain network attributes in accurately distinguishing individuals across multiple datasets.

Table 5.1 Classification results of four datasets (%)

k	BCI data	DRI data	NMK data	Motor imagery dataset
1	100.0	100.0	94.4	100.0
2	98.6	99.0	96.3	100.0
3	100.0	100.0	95.6	100.0
4	97.2	99.0	95.0	100.0
5	97.2	97.9	94.4	98.2
6	98.6	100.0	98.1	96.4
7	100.0	98.9	96.3	100.0
8	100.0	98.9	94.4	99.1
9	100.0	96.9	95.6	100.0
10	100.0	100.0	96.3	100.0
Average	99.2 ± 0.01	99.1 ± 0.01	95.6 ± 0.01	99.4 ± 0.01

Table 5.2 Comparison of our method and AR+PSD and Nguyen's method

Features	Datasets	Accuracy
Brain network attributes	BCI data	0.992
	DRI data	0.991
	NMK data	0.956
AR+PSD	BCI data	0.689
	DRI data	0.382
	NMK data	0.224
Nguyen's	BCI data	0.46

To further validate the discriminative power of the proposed features, a comparison is performed against fixed features comprised of Autoregressive (AR) parameters and Power Spectral Density (PSD), as proposed in prior studies [5, 25, 26]. The classification performance using AR and PSD features is summarized in Table 5.2.

As shown in Table 5.2, the accuracies obtained using AR parameters are consistently lower than those achieved with brain network attributes. For all three datasets analyzed, brain network attributes demonstrate superior identification performance. Notably, for the BCI dataset, the fixed parameters achieve slightly higher accuracy compared to the other two datasets, but they remain inferior to the results obtained using brain network attributes.

Additionally, the performance of the proposed method is compared with an approach based on mel-frequency cepstral coefficients (MFCCs), as reported in [27]. In that study, an identification accuracy of 0.46 was achieved using the Graz A 2008 dataset (referred to as BCI data). In contrast, the proposed method significantly outperforms this baseline, achieving an accuracy exceeding 0.99. Unlike the MFCC-based approach, which demonstrates limited applicability, the proposed method provides robust and highly accurate personal identification across diverse datasets.

5.3.3 Effect of Four EEG Bands

EEG signals are categorized into four frequency bands: theta (4–8 Hz), alpha (8–13 Hz), beta (13–30 Hz), and gamma (30–40 Hz). The intensity of these signals varies with mental states, making certain bands more effective for specific tasks. To evaluate the performance of brain network attributes in personal identification across these bands, a comparison is performed. Each dataset undergoes 10 rounds of twofold cross validation, and the mean classification accuracies are visualized in Fig. 5.9.

As depicted in Fig. 5.9, the classification accuracies in the beta, alpha, and gamma bands are consistently higher than those in the theta band across all datasets. An

5.3 Brain Fingerprint Identification with Brain Network

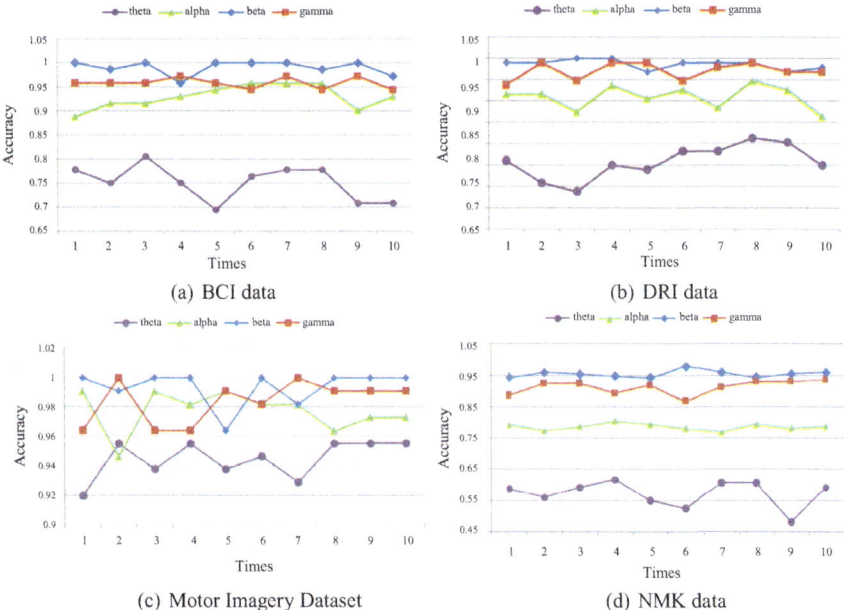

Fig. 5.9 Four bands classification accuracies for four datasets

exception occurs in the Motor Imagery Dataset, where the theta band exhibits competitive accuracy in a single instance. These results indicate that the beta and gamma bands are the most effective for brain fingerprint identification using brain network attributes. This can be attributed to the stronger connectivity patterns observed in these bands, which provide better discrimination of individual characteristics.

5.3.4 Effect of Combined Attributes

To explore the impact of combined attributes on classification performance, experiments are conducted using both single attributes and a combination of attributes across all datasets. The results, summarized in Table 5.3, demonstrate the superiority of combining attributes in brain fingerprint identification tasks.

Table 5.3 Classification results of different features (%)

Attributes	BCI data	DRI data	NMK data	Motor imagery dataset
D	98.0	98.6	94.5	99.1
D+Eg+C	**99.2**	**99.1**	**95.6**	**99.4**

D: Degree of nodes Eg: Global efficiency C: Clustering coefficient

Table 5.3 reveals that the classification accuracy for single attributes, such as the degree of nodes, is generally good but consistently lower than that achieved with combined attributes. Incorporating global efficiency and clustering coefficient into the feature vector significantly enhances performance, underscoring the complementary nature of these attributes in capturing individual-specific patterns.

For the datasets analyzed, the subject counts are as follows: 9 for the BCI dataset, 14 for the Motor Imagery Dataset, 20 for the NMK dataset, and 12 for the DRI dataset. Notably, the NMK dataset has the largest subject count but exhibits the lowest classification accuracy. This suggests that larger sample sizes increase the likelihood of overlapping attribute values among individuals, which can reduce discriminative power. Despite this, the combination of attributes demonstrates robust performance across varying dataset sizes, confirming its effectiveness in improving classification accuracy.

5.4 Conclusion

This chapter introduces an EEG-based method for estimating cognitive phenotypes in brain fingerprint identification, leveraging features derived from the brain's functional networks. Unlike traditional approaches, this method constructs feature vectors based on brain network attributes, offering a novel perspective for personal identification.

The proposed approach is evaluated across four datasets, each representing distinct cognitive tasks, and achieves classification accuracies exceeding 0.95 in all cases. Remarkably, even in the challenging fatigue driving task—characterized by complex cognitive demands—the method demonstrates robust performance. These findings underscore the uniqueness of individual brain network attributes as reliable indicators for identification.

Moreover, the analysis reveals that EEG signals in the beta and gamma frequency bands carry significantly more informative features when interpreted through brain network analysis. This highlights the importance of frequency-specific investigation in enhancing identification accuracy.

In summary, the results confirm that brain network attributes are promising and effective features for brain fingerprint identification, paving the way for further exploration of EEG-based cognitive profiling in practical applications.

References

1. Hebb DO (2013) The organization of behavior; A neuropsychological theory. Chapman & Hall, Wiley
2. Fries P (2005) A mechanism for cognitive dynamics: neuronal communication through neuronal coherence. Trends Cogn Sci 9(10):474
3. Ling W, Li Y, Yang X, Xue Q, Wang Y (2015) Altered characteristic of brain networks in mild cognitive impairment during a selective attention task: an EEG study. Int J Psychophysiol 98(1):8–16
4. Jamal W, Das S, Maharatna K, Pan I, Kuyucu D (2015) Brain connectivity analysis from EEG signals using stable phase-synchronized states during face perception tasks. Physica A Stat Mech Its Appl 434:273–295
5. Fei Su, Xia Liwen, Cai Anni, Ma Junshui (2010) Evaluation of recording factors in EEG-based personal identification: a vital step in real implementations. In: IEEE international conference on systems, man and cybernetics, Istanbul, Turkey 10–13 October, pp 3861–3866
6. Rubinov M, Sporns O (2009) Complex network measures of brain connectivity: uses and interpretations. Neuroimage 52(3):1059–1069
7. Mcfarland DJ, Mccane LM, David SV, Wolpaw JR (1997) Spatial filter selection for EEG-based communication. Electroencephalogr Clin Neurophysiol 103(3):386–394
8. Van Quyen M Le, Foucher J, Lachaux J, Rodriguez E, Lutz A, Martinerie J, Varela FJ (2001) Comparison of Hilbert transform and wavelet methods for the analysis of neuronal synchrony. J Neurosci Methods 111(2):83–98
9. Rosenblum MG, Pikovsky AS, Kurths J (1996) Phase synchronization of chaotic oscillators. Phys Rev Lett 76(11):1804
10. Kong W, Zhou Z, Jiang B, Babiloni F, Borghini G (2017) Assessment of driving fatigue based on intra/inter-region phase synchronization. Neurocomputing 219:474–482
11. Rosenblum MG, Pikovsky AS, Kurths J (2012) Synchronization approach to analysis of biological systems. Fluctuation Noise Lett 04(1):L53–L62
12. Bullmore E, Sporns O (2009) Complex brain networks: graph theoretical analysis of structural and functional systems. Nat Rev Neurosci 10(3):186–198
13. Park HJ, Friston K (2013) Structural and functional brain networks: from connections to cognition. Science 342(6158):1238411
14. Stam CJ (2009) From synchronisation to networks: assessment of functional connectivity in the brain. Springer, New York
15. Sakkalis V, Oikonomou T, Tsiaras V, Tollis I (2015) Graph-theoretic indices of evaluating brain network synchronization: application in an alcoholism paradigm. Neuromethods 91:159–169
16. Chavez M, Valencia M, Latora V, Martinerie J (2010) Complex networks: new trends for the analysis of brain connectivity. Int J Bifurcat Chaos 20(6):1677–1686
17. Lei G, Wang Yao Yu, Hongli YN, Ying L (2014) Study of brain functional network based on sample entropy of EEG under magnetic stimulation at pc6 acupoint. Bio-med Mater Eng 24(1):1063–9
18. Boccaletti S, Latora V, Moreno Y, Chavez M, Hwang DU (2006) Complex networks: structure and dynamics. Phys Rep 424(4–5):175–308
19. Latora V, Marchiori M (2001) Efficient behavior of small-world networks. Phys Rev Lett 87(19):198701
20. Saramäki J, Kivelä M, Onnela J-P, Kaski K, Kertész J (2007) Generalizations of the clustering coefficient to weighted complex networks. Phys. Rev. E, Stat, Nonlinear, Soft Matter Phys 75(2 Pt 2):027105
21. Onnela JP, Saramäki J, Kertész J, Kaski K (2005) Intensity and coherence of motifs in weighted complex networks. Phys Rev E 71(6 Pt 2):065103
22. Ye J, Janardan R, Li Q (2004) Two-dimensional linear discriminant analysis. Photogram Eng Remote Sens 5(6):1431–1441
23. Peng Y, Bao-Liang L (2017) Discriminative extreme learning machine with supervised sparsity preserving for image classification. Neurocomputing 261:242–252

24. Kim TK, Kim H, Hwang W, Kee SC (2003) Face description based on decomposition and combining of a facial space with LDA. In: International conference on image processing, 2003. ICIP 2003. Proceedings, pp 877–880
25. Poulos M, Rangoussi M, Alexandris N (1999) Neural network based person identification using EEG features. In: IEEE international conference on acoustics, speech, and signal processing, pp 1117–1120
26. Hema CR, Paulraj MP, Kaur H (2009) Brain signatures: a modality for biometric authentication. In: International conference on electronic design, pp 1–4
27. Nguyen P, Tran D, Huang X, Sharma D (2012) A proposed feature extraction method for EEG-based person identification. In: Proceedings of the 2012 international conference on artificial intelligence

Open Access This chapter is licensed under the terms of the Creative Commons Attribution-NonCommercial-NoDerivatives 4.0 International License (http://creativecommons.org/licenses/by-nc-nd/4.0/), which permits any noncommercial use, sharing, distribution and reproduction in any medium or format, as long as you give appropriate credit to the original author(s) and the source, provide a link to the Creative Commons license and indicate if you modified the licensed material. You do not have permission under this license to share adapted material derived from this chapter or parts of it.

The images or other third party material in this chapter are included in the chapter's Creative Commons license, unless indicated otherwise in a credit line to the material. If material is not included in the chapter's Creative Commons license and your intended use is not permitted by statutory regulation or exceeds the permitted use, you will need to obtain permission directly from the copyright holder.

Chapter 6
Multi-task and Single-Session Brain Fingerprint Identification with Low-Rank and Matrix Decomposition

Abstract EEG-based brain fingerprint identification usually requires participators to complete a particular task under external stimuli, such as recognition based on visual-evoked potentials (VEP) and resting potential (RP). These paradigms often fail to generalize due to their dependency on task-specific brain responses, which restricts their applicability in dynamic real-world environments. This chapter proposes a fast task-free brain fingerprint identification method based on Low-Rank and Matrix Decomposition (LRMD). Task-related EEG signals are conceptually divided into two components: the background EEG (BEEG) and the residue EEG (REEG). The BEEG encapsulates an individual's intrinsic and unique brain print features that remain stable over time. Moreover, only a subset of these brainprint features is sufficient to characterize an individual's identity. Consequently, the BEEG can be represented by a compact set of basis vectors in its feature space, exhibiting low-rank characteristics. The LRMD framework effectively isolates identity-related components by decomposing raw EEG signals into meaningful subspaces while filtering out noise and irrelevant features. This capability allows for accurate recognition even without requiring participants to engage in predefined tasks or external stimuli. Extensive experiments are conducted on three public EEG datasets and a self-collected multi-task EEG dataset, demonstrating outstanding performance across varying low-rank configurations and diverse time scales. These results indicate that the proposed method is robust and task-agnostic, further validating its potential for practical applications. The best results can reach accuracy above 99.90%, highlighting its potential for widespread application in practical scenarios such as authentication and secure access systems.

6.1 Introduction

EEG-based brain fingerprint identification garners significant attention due to its potential for secure and personalized authentication. Current methods, classified into four categories: resting potentials, visual-evoked potentials (VEP), movement imagination, and event-related potentials (ERP), rely heavily on task-specific EEG responses. However, these approaches face notable limitations in real-world scenarios.

Identification methods based on VEP [1, 2] require participants to complete additional cognitive tasks with visual stimulation equipment, making them unsuitable for individuals with impaired vision. Similarly, RP-based methods [3] require highly relaxed states, which are challenging to maintain in uncontrolled environments. Furthermore, task-specific approaches are inherently limited to single-task paradigms, restricting their generalizability.

Supawich Puengdang et al. [4] combine steady-state visual-evoked potentials (SSVEP) and ERP features with a Long Short-Term Memory (LSTM) network for EEG authentication. Specifically, they collect raw SSVEP EEG data from 20 human subjects and ERP EEG data elicited through Snodgrass-Vanderwart image stimulation. The LSTM architecture processes individual EEG data, enabling accurate prediction and verification. This study demonstrates the feasibility of multi-task brain fingerprint identification. Moreover, Bengson et al. [5] highlight the existence of spontaneous brain activity patterns across the entire brain. Complementarily, low-rank matrix decomposition methods are employed to locate brain signal sources. Gramfort et al. [6] propose a time-frequency mixed-norm estimate for high-accuracy brain signal source localization, confirming the decomposability of EEG signals and the low-rank characteristics of background EEG.

Building upon these findings, this chapter introduces a task-free brain fingerprint identification method leveraging the Low-Rank and Matrix Decomposition (LRMD) model. In this approach, task-related EEG signals are divided into Background EEG (BEEG) and Residue EEG (REEG). The BEEG represents intrinsic and unique brainprint features with low-rank properties, while the REEG comprises noise and task-evoked cortical signals. Notably, the BEEG is less sensitive to task-related variations, exhibiting stability over time.

Leveraging low-rank decomposition in EEG-based biometrics is important because it can simultaneously address noise reduction and feature extraction. Low-rank models are particularly effective for isolating stable patterns, which are essential for reliable identity verification. By separating stable brainprint characteristics from dynamic, task-induced activities, LRMD not only improves classification accuracy but also ensures robustness across varying conditions.

Furthermore, advances in computational efficiency make such models viable for practical deployment. Modern hardware and algorithmic optimizations allow real-time processing of EEG signals, making task-free biometrics a feasible option for secure authentication in diverse applications.

The proposed LRMD model incorporates an efficient kernel function for rapid BEEG extraction and combines sparse representation for high-precision classification. By employing a smaller dictionary, this method achieves superior spatial efficiency compared to traditional sparse representation algorithms [7]. Additionally, replacing the Gaussian kernel with a quadratic rational kernel [8] ensures classification accuracy while significantly reducing computational complexity, offering a practical and scalable solution for real-world applications.

6.2 Low-Rank and Matrix Decomposition

In this section, the entire process of the proposed approach for brain fingerprint identification is outlined. Initially, the raw EEG data is transformed into an EEG spectrogram using the Short-Time Fourier Transform (STFT). The resulting spectrogram is then partitioned into training and testing datasets. Then, an efficient LRMD algorithm, GoDec+, is applied, combined with a fast Rational Quadratic (RQ) kernel, to perform BEEG feature extraction on the training dataset. Subsequently, the ensemble subspace, which represents the underlying structure of all samples, is constructed. Upon receiving a test sample, the optimal reconstruction coefficient is determined using the Minimum Cross-Entropy Criterion (MCC). Finally, classification is achieved by computing the cross-entropy between the test sample and its corresponding reconstructed sample. A schematic overview of the entire process is provided in Fig. 6.1. The process begins with the transformation of the original EEG data into an EEG spectrogram via STFT. In the figure, In the figure, a single column of the matrix $\mathbf{S}$ represents one sample, and each piece of data consists of all samples associated with a particular subject. The raw EEG data is then randomly partitioned into two sets: the training data set, denoted as $\mathbf{S}_1$, and the testing data set, denoted as $\mathbf{S}_2$. The primary goal of the training phase is to construct the ensemble subspace, which can be thought of as a dictionary representing the features of the EEG data. To achieve this, $\mathbf{S}_1$ is decomposed into two components, $\mathbf{R}$ and $\mathbf{B}$. This decomposition is followed by QR decomposition to generate the ensemble subspace $\mathbf{D}^{m \times kr}$ from

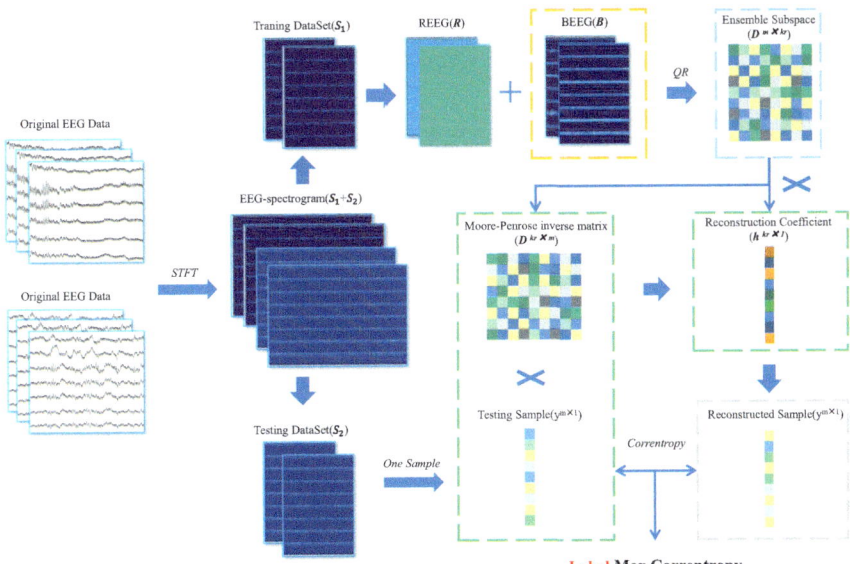

Fig. 6.1 The whole process of brain fingerprint identification based on LRMD model

the r-rank matrix $\mathbf{B}^{m \times n}$, where m denotes the number of samples and kr represents the reduced rank of the matrix. The calculation of the Moore-Penrose inverse matrix $\mathbf{M}^{kr \times m}$ is detailed in Sect. 6.2.4.

During the testing phase, the reconstruction coefficient vector $\mathbf{h}$ is computed using an m-dimensional test sample $\mathbf{y}$ and the matrix $\mathbf{M}^{kr \times m}$. The detailed procedure is described in Algorithm 6.1. The corresponding new sample $\hat{\mathbf{y}}$, which resides in the ensemble subspace $\mathbf{D}^{m \times kr}$, is efficiently located through the kr-dimensional vector $\mathbf{h}$. To refine the reconstructed sample, a mask function is applied to the vector $\mathbf{h}$, allowing the generation of k reconstructed samples, as presented in Sect. 6.2.5. Finally, classification is performed by calculating the cross-entropy between the test sample and the reconstructed samples. The test sample is assigned to the class where the cross-entropy is maximized.

Algorithm 6.1 Optimal reconstruction coefficient generation

Require: $\mathbf{y}, \mathbf{D} \in \mathbb{R}^{m \times kr}, \alpha_2, \beta_2, \epsilon = 10^{-7}, \text{Iter}_2$.
Ensure: h
1: Initialize: $\mathbf{e} = \mathbf{0}, \mathbf{h}_{(0)} = \mathbf{0}, t := 0$.
2: $\mathbf{M} = (\mathbf{D}^T \mathbf{D} + \lambda \mathbf{I})^{-1} \mathbf{D}^T$;
3: **while** $t < \text{Iter}_2$ or $\|\mathbf{h}_{(t)} - \mathbf{h}_{(t-1)}\|_2^2 / \|\mathbf{h}_{(t-1)}\|_2^2 > \epsilon$ **do**
4: $\quad t : t + 1$;
5: $\quad \hat{\mathbf{y}} = \mathbf{y} - \mathbf{e}, \mathbf{h}_{(t)} = \mathbf{M}\hat{\mathbf{y}}$;
6: $\quad \boldsymbol{\tau} = \mathbf{y} - \mathbf{D}\mathbf{h}_{(t)}, \mathbf{e} = \boldsymbol{\tau} - \boldsymbol{\tau} \circ k_{\alpha_2, \beta_2}(\boldsymbol{\tau})$;
7: **end while**

6.2.1 Domain Transformation of EEG Signals

In general, it is challenging to extract the brainprint features from non-stationary EEG signals in the time domain [6, 9]. Non-stationary signals, characterized by their varying statistical properties over time, require specialized techniques for analysis. Traditional methods, such as Short-Time Fourier Transform (STFT) [10] and Wavelet Decomposition (WD) [11], have been widely adopted in industry for processing such signals. For WD, a pair of wavelet basis functions must typically be designed beforehand to appropriately capture the signal's features at different scales. However, to optimize processing speed, STFT is often preferred for segmenting the EEG signal. STFT divides the c-channel EEG signal $\mathbf{X}$ into l segments, each of a predefined time length. This division is performed in a way that ensures an even distribution of time across the segments, which can be mathematically represented as follows:

6.2 Low-Rank and Matrix Decomposition

$$\mathbf{X} = \begin{bmatrix} \mathbf{x}_{11} & \cdots & \mathbf{x}_{1c} \\ \vdots & \ddots & \vdots \\ \mathbf{x}_{l1} & \cdots & \mathbf{x}_{lc} \end{bmatrix}, \tag{6.1}$$

where $\mathbf{x}_{ij}$ represents the sub-signal of channel j in segment i. Subsequently, the signal $\mathbf{x}ij$ is transformed into the frequency domain using the Fast Fourier Transform (FFT) method [12], where the signal becomes more stable and easier to analyze. This transformation is represented as $\mathbf{s}ij = f(\mathbf{x}_{ij})$, where $f(\cdot)$ denotes the FFT function. Following the application of STFT, the signal from each channel is converted from a time series into a frequency spectrum. This frequency spectrum captures the signal's properties at different frequencies and time intervals, revealing crucial insights into the signal's characteristics. Given the inherent insensitivity of brainprint features to time, the transformed signal in each segment is treated as a distinct sample representing the subject. Intuitively, a shorter time segment results in more samples being available for analysis, which allows the LRMD model to accumulate a larger training dataset. On the other hand, larger time segments provide more detailed information, allowing the model to uncover richer brainprint features, which facilitates a more comprehensive training process. The challenge, therefore, is to strike an optimal balance–one that ensures the data pool is sufficiently large while maintaining the richness and complexity of each individual sample. All subsequent algorithms in this study rely on EEG spectrograms, which are generated through this process, as illustrated in Fig. 6.2.

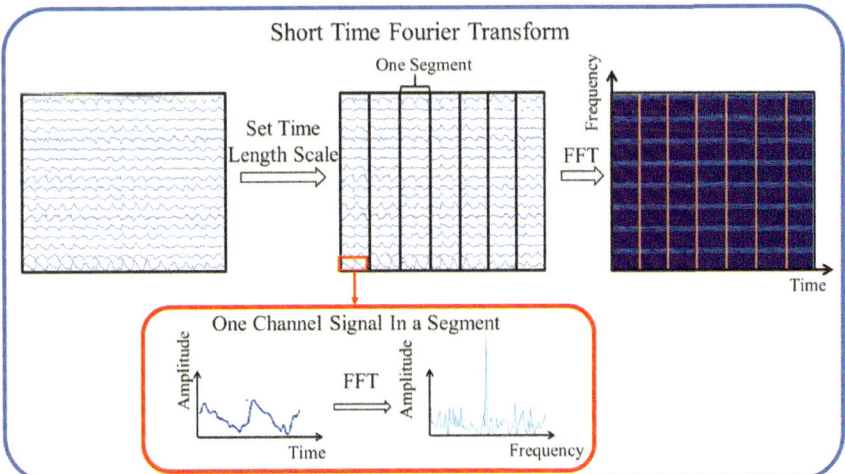

Fig. 6.2 The entire process of applying the Short-Time Fourier Transform (STFT) is as follows: First, the original EEG data is uniformly divided into multiple segments based on a predefined time length scale. Then, the Fast Fourier Transform (FFT) method is applied to each segment, transforming the signal of one channel from the time domain into the frequency domain. As a result, the 1D EEG signal is converted into a 2D frequency spectrum, which is more suitable for matrix decomposition techniques

It is important to note that the STFT applies the FFT individually to the data within each segment, without any interaction between the segments. This means that each segment is processed independently, and as a result, each sample is generated independently of the others. Consequently, the training samples and testing samples remain independent from one another, ensuring that the training and evaluation phases are conducted on separate, non-interfering data sets.

6.2.2 LRMD Model-Based Background EEG Extraction

6.2.2.1 LRMD Model

Based on our assumption that an EEG signal related to a specific task can be effectively partitioned into two distinct components, the EEG spectrogram **S** can be decomposed as follows:

$$\mathbf{S} = \mathbf{B} + \mathbf{R}, \tag{6.2}$$

where **B** represents a low-rank matrix that encapsulates the Brain-EEG (BEEG) component, while **R** corresponds to the residual EEG, which consists of the task-evoked EEG and random noise. Although the model is simple, it is both powerful and effective in separating the task-related information from the residual noise. To further enhance the model's performance, the rank of the matrix **B** is constrained, and the reconstruction error is minimized.

6.2.2.2 GoDec+ Algorithm

The rank of a matrix is a measure of its two-dimensional sparsity [13], and numerous algorithms have been developed in recent years to optimize the LRMD model [14]. A notable contribution by Ebadi and Izquierdo [15] introduced a novel Singular Value Decomposition (SVD)-free algorithm specifically designed for the rank-1 background case. Among the most widely recognized techniques are the Alternating Direction Method (ADM) [16] and the GoDec algorithm [17]. The ADM algorithm applies sparse constraints through the l_1 norm and enforces low-rank constraints using the nuclear norm, which allows for fast matrix decomposition. On the other hand, the GoDec algorithm enhances the decomposition speed and robustness by incorporating random bilateral projection techniques. Furthermore, Guo et al. introduced the MCC into the GoDec algorithm, which significantly improved its performance. This enhancement led to the development of the anti-jamming GoDec+ algorithm [8], which is faster and more resilient to noise interference.

To extract the BEEG features, the GoDec+ algorithm is employed. GoDec+ is a fast and robust low-rank matrix decomposition algorithm that requires the use of a kernel function to enhance its performance. When combined with the LRMD model

6.2 Low-Rank and Matrix Decomposition

outlined earlier, the objective function and the constraints for the GoDec+ algorithm can be formulated as follows:

$$\min_{\mathbf{B},\mathbf{E}} \|\mathbf{S} - \mathbf{B} - \mathbf{E}\|_F^2 + \varphi_v(\mathbf{E}), \quad \text{s.t. rank}(\mathbf{B}) \leq r, \tag{6.3}$$

where $\mathbf{E}$ is an auxiliary variable introduced by the Hinge-Quadratic (HQ) method [18], and r denotes the upper bound on the rank of matrix $\mathbf{B}$. Additionally, $\varphi_v(\mathbf{E})$ is defined as $\varphi_v(\mathbf{E}) = \sum_{i=1}^{m} \sum_{j=1}^{n} \varphi(\mathbf{E}i, j)$, where $\varphi(\cdot)$ is the dual function corresponding to the kernel function $k(\cdot)$. Following Guo et al. [8], we adopt the Gaussian kernel function $k(\cdot) = g\sigma(x) = \exp(-x^2/\sigma^2)$.

The optimization problem in Eq. (6.3) can be solved using the Alternating Least Squares (ALS) method. This results in the following two updated equations:

$$\begin{cases} \mathbf{B}_{(t)} = \arg\min_{rank(\mathbf{B}) \leq r} \|\mathbf{S} - \mathbf{B} - \mathbf{E}_{(t-1)}\|_F^2 + \varphi_v(\mathbf{E}_{(t-1)}) \\ \mathbf{E}_{(t)} = \arg\min_{\mathbf{E}} \|\mathbf{S} - \mathbf{B}_{(t)} - \mathbf{E}\|_F^2 + \varphi_v(\mathbf{E}). \end{cases} \tag{6.4}$$

When $\mathbf{B}$ is fixed, $\mathbf{E}$ can be updated using the HQ method, which yields $\mathbf{E} = \mathbf{R} - \mathbf{R} \circ k(\mathbf{R})$, where $\circ$ denotes the Hadamard product. On the other hand, when $\mathbf{E}$ is fixed, $\mathbf{B}$ can be computed by solving the first equation in Eq. (6.4) using the GreB paradigm [19].

Combined with the bilateral projection method from Zhou and Tao [19], an iterative QR decomposition-based update method can be formulated as:

$$\begin{cases} \mathbf{U}_{(k)} = \mathbf{Q}, \quad \mathbf{SV}_{k-1}^T = \mathbf{QW}, \\ \mathbf{V}_{(k)} = \mathbf{Q}^T \mathbf{S}, \end{cases} \tag{6.5}$$

where $\mathbf{QW}$ represents the QR decomposition of $\mathbf{SV}_{k-1}^T$. The QR decomposition process transforms the original matrix into an upper triangular matrix through an orthogonal similarity transformation, simplifying the computation of the eigenvalues. (To avoid confusion of the character $\mathbf{R}$ in Eq. (6.2), $\mathbf{QW}$ is used to denote QR decomposition.)

6.2.2.3 Rational Quadratic Kernel Function

In this section, the rational quadratic (RQ) kernel [20] is employed instead of the Gaussian kernel. The RQ kernel is defined as:

$$k_{\alpha,\beta}(x) = \left(1 + \frac{\alpha}{\beta}x^2\right)^{-\beta}, \tag{6.6}$$

Algorithm 6.2 GoDec+

Require: $\mathbf{S} \in \mathbb{R}^{m \times n}, r, q, \alpha_1, \beta_1, \epsilon = 10^{-7}, \text{Iter}_1$.
Ensure: B
1: Initialize: $\mathbf{E}_{(0)} = \mathbf{0}, \mathbf{B}_{(0)} = \mathbf{S}, t := 0$.
2: Generated standard Gaussian matrix $\mathbf{A} \in \mathbb{R}^{n \times r}$;
3: **while** $t < \text{Iter}_1$ or $\|\mathbf{B}_{(t)} - \mathbf{B}_{(t-1)}\|_F^2 / \|\mathbf{B}_{(t-1)}\|_F^2 > \epsilon$ **do**
4: $\quad t : t + 1$;
5: $\quad \hat{\mathbf{S}} = \mathbf{S} - \mathbf{E}_{(t-1)}$;
6: $\quad$ **for** $i = 1$ to q **do**
7: $\quad\quad \mathbf{Y}_1 = \mathbf{A}^T \hat{\mathbf{S}}, \mathbf{Y}_2 = \hat{\mathbf{S}} \mathbf{Y}_1^T$;
8: $\quad$ **end for**
9: $\quad \mathbf{Y}_2 = \mathbf{QW}, \mathbf{A} = \mathbf{Q}$;
10: $\quad \mathbf{B}_{(t)} = \mathbf{Q}\mathbf{Q}^T \hat{\mathbf{S}}, \mathbf{R}_{(t)} = \mathbf{S} - \mathbf{B}_{(t)}, \mathbf{E}_{(t)} = \mathbf{R}_{(t)} - \mathbf{R}_{(t)} \circ k_{\alpha_1, \beta_1}(\mathbf{R}_{(t)})$;
11: **end while**

where α is a width parameter that controls the spread of the kernel, and β is a weighting parameter that governs the relative importance of large-scale and small-scale variations. Notably, when the value of β is large, the RQ kernel becomes similar to the Gaussian kernel.

Theorem 6.1 *When $\beta \to \infty$, the RQ kernel is identical to the Gaussian kernel.*

Proof The variables x and α in Eq. (6.6) can be treated as constants when considering the limit as $\beta \to \infty$. We transform (6.6) into the following expression:

$$\lim_{\beta \to \infty} \left(1 + \frac{\alpha}{\beta} x^2\right)^{-\beta} = \lim_{\beta \to \infty} \left(\left(1 + \frac{\alpha}{\beta} x^2\right)^{\frac{\beta}{\alpha x^2}}\right)^{-\alpha x^2} \quad (6.7)$$

where $\lim_{\beta \to \infty} (1 + \frac{\alpha}{\beta} x^2)^{\frac{\beta}{\alpha x^2}}$ is recognized as the natural exponential constant e [21]. Therefore, Eq. (6.7) simplifies to:

$$\lim_{\beta \to \infty} \left(1 + \frac{\alpha}{\beta} x^2\right)^{-\beta} = \lim_{\beta \to \infty} e^{-\alpha x^2}$$
$$= e^{-\alpha x^2} \quad (6.8)$$

which is precisely the form of a Gaussian kernel function. $\square$

Therefore, when β is set to a specific value, the RQ kernel serves as a reasonable approximation of the Gaussian kernel. This property can be used to significantly reduce the computational load while maintaining the accuracy of the classification.

Based on the above solutions and the acceleration scheme in Zhou and Tao [17], the improved GoDec+ algorithm is summarized in Algorithm 6.2. In this algorithm, q is the acceleration coefficient, ϵ represents the error tolerance, and Iter_1 denotes the maximum number of iterations.

6.2.3 Ensemble Subspace Construction

Low-rank matrix decomposition requires all the samples of a subject for cross-correlation analysis. Therefore, it is inevitable that testing data may become involved in the training process. However, by incorporating sparse representation [7], samples that do not participate in the low-rank matrix decomposition can still be successfully classified. This allows for a clear separation between the testing and training datasets. The BEEG matrix **B** can be viewed as a dictionary, enabling the classification of test samples by identifying the class to which they belong. This process is commonly referred to as dictionary learning. The basis of the matrix **B** can be used to regenerate a smaller dictionary that serves as a substitute for **B**, thereby accelerating the processes in Algorithms 6.1 and 6.3.

Algorithm 6.3 MCC-based classification

Require: $\mathbf{y}, \mathbf{D} \in \mathbb{R}^{m \times kr}, \mathbf{h}$.
Ensure: identity($\mathbf{y}$)
1: Compute the correntropy for each subject
   $C_i(\mathbf{y}) = \sum_{j=1}^{m} k_{\alpha,\beta}[(\mathbf{y} - \mathbf{D}\delta_i(\mathbf{h}))_j], \forall i = 1, ..., k$.
2: Identify($\mathbf{y}$) = $\arg \max_i C_i(\mathbf{y})$

Let $\mathbf{B}^i$ represent the BEEG of subject i. Through QR decomposition, a lightweight dictionary **D** for k subjects can be constructed as follows:

$$\begin{cases} \mathbf{B}^i = \mathbf{Q}^i \mathbf{W}^i, \\ \mathbf{D}^i = \mathbf{Q}^i_{1:r}, \\ \mathbf{D} = [\mathbf{D}^1, ..., \mathbf{D}^k], \end{cases} \quad (6.9)$$

where $\mathbf{Q}^i_{1:r}$ denotes the first r columns of $\mathbf{Q}^i$, and r is the rank of the BEEG data matrix **B**.

6.2.4 Reconstruction Coefficient Generation

In dictionary learning, the reconstruction coefficient acts as an index, allowing for quick retrieval of the corresponding object in the dictionary. When a test sample $\mathbf{y}$ is presented, the optimal reconstruction coefficient **h** can be computed by solving the following optimization problem:

$$\min_{\mathbf{h}} \sum_{j=1}^{m} \{1 - k_{\alpha,\beta}[(\mathbf{y} - \mathbf{Dh})_j]\} + \lambda \|\mathbf{h}\|_2^2, \quad (6.10)$$

where λ is the trade-off parameter. Following [22], an error vector $\mathbf{e}$ is introduced, yielding the following optimization problem:

$$\min_{\mathbf{h},\mathbf{e}} \|\mathbf{y} - \mathbf{Dh} - \mathbf{e}\|_2^2 + \sum_{j=1}^{m}[\varphi(\mathbf{e}_j)] + \lambda \|\mathbf{h}\|_2^2. \quad (6.11)$$

This optimization problem in Eq. (6.11) can be solved using the ALS method.

When $\mathbf{h}$ is settled, we have $\mathbf{e} = \boldsymbol{\tau} - \boldsymbol{\tau} \circ k_{\alpha,\beta}(\boldsymbol{\tau})$, where $\boldsymbol{\tau} = \mathbf{y} - \mathbf{Dh}$. When $\mathbf{e}$ is fixed, the reconstruction coefficient $\mathbf{h}$ can be calculated by: $\mathbf{h} = (\mathbf{D}^T\mathbf{D} + \lambda\mathbf{I})^{-1}\mathbf{D}^T\hat{\mathbf{y}}$, where $(\mathbf{D}^T\mathbf{D} + \lambda\mathbf{I})^{-1}\mathbf{D}^T$ is the Moore-Penrose pseudoinverse of $\mathbf{D}$. Although computing the matrix inverse may be computationally expensive, it only needs to be calculated once before the algorithm starts.

6.2.5 MCC-Based Classification

The reconstruction coefficient $\mathbf{h}$ is a very long and sparse vector, and to improve computational efficiency, a mask is introduced. A mask function $\delta_i(\cdot) : \mathbb{R}^n \to \mathbb{R}^n$ is defined, which exposes only the part of $\mathbf{h}$ corresponding to subject i, while the other components are set to zero. Consequently, the reconstructed sample can be computed as: $\tilde{\mathbf{y}} = \mathbf{D}\delta_i(\mathbf{h})$. Next, the cross-entropy between $\mathbf{y}$ and $\tilde{\mathbf{y}}$ is calculated using the MCC approach. Finally, classification is performed as outlined in Algorithm 6.3.

At this point, the detailed description of the entire brain recognition process is concluded.

6.3 Brain Fingerprint Identification with LRMD

In this section, the above algorithm is applied to extensive experiments using three public datasets: the BCI Graz Dataset A [23], the P300 EEG Dataset [24], and the SJTU Emotion EEG Dataset (SEED) [25], as well as a self-collected Multi-task EEG Dataset (MTED). A brief overview of all datasets is provided in Table 6.1.

Table 6.1 The description of the datasets

Name	Channel	Subject	Sample frequency (Hz)
SEED	62	15	200
MTED	62	15	200
BCI Graz dataset A	22	9	250
P300 EEG dataset	8	8	256

6.3 Brain Fingerprint Identification with LRMD

Currently, there are very few databases that are collected under various tasks with the same subjects. Therefore, an experimental paradigm is designed, combining a variety of task-evoked EEG data to create a unified set of EEG data suitable for the proposed task-free brain fingerprint identification approach. The impact of different ranks and time length scales on recognition accuracy is then analyzed and compared. Additionally, using the BCI Graz Dataset A, the time consumption of the Gaussian kernel and the RQ kernel is compared. The code for the experiments is available for download at https://github.com/kxh001/Task-Independent-EEG-Identification.

6.3.1 Implementation Details

For the RQ kernel, the width variable, weighting variable, and accelerating coefficient are set as follows: $\alpha_1 = \alpha_2 = 0.0001$, $\beta_1 = \beta_2 = 1$, and $q = 1$. For the Gaussian kernel, the parameter σ is set to 0.0001. In each algorithm, the maximum number of iterations is set as $\text{Iter}_1 = \text{Iter}_2 = 100$. The rank of the BEEG data matrix varies from 1 to 10, and the time length scale for STFT is set to 1, 5, 10, and 20 s sequentially. Each experiment is repeated 10 times for each rank and time length scale. In all experiments, the total number of samples per subject is no <40, with the number of samples increasing proportionally with the time length scale. For instance, when the time length scale is 20 s and the number of samples is 40, this increases to 800 samples when the time length scale is reduced to 1 s. The ratio of training samples to testing samples is always 1:1 in our experiments.

6.3.2 Performances of Multi-task and Single-Session Brain Fingerprint Identification

Both Figs. 6.3 and 6.4 demonstrate the superiority of the fast RQ kernel-based GoDec+ algorithm, highlighting that BEEG indeed contains a low-rank intrinsic brainprint. For single-task datasets such as the BCI Graz dataset A and the P300 dataset, our algorithm achieves an accuracy >95%. For the MTED dataset, when the rank is set to 8 and the time length scale is 1 s, the accuracy reaches 99.88%. This shows that our approach overcomes the limitation of brain fingerprint identification based solely on single-task data. Furthermore, when applied to the SEED dataset, the accuracy exceeds 99%, indicating that the approach not only adapts to task diversity but also tolerates emotional variability. Particularly, as seen in Fig. 6.4b and c, the model proves to be highly robust, bringing brain fingerprint identification closer to real-world applications.

From Fig. 6.4, it is evident that as the rank of BEEG increases beyond 3, the recognition accuracy gradually stabilizes. In some cases, when the rank exceeds 5, the recognition accuracy slightly decreases, suggesting that brainprint features are

Fig. 6.3 Recognition accuracy (%) on BCI Graz dataset A

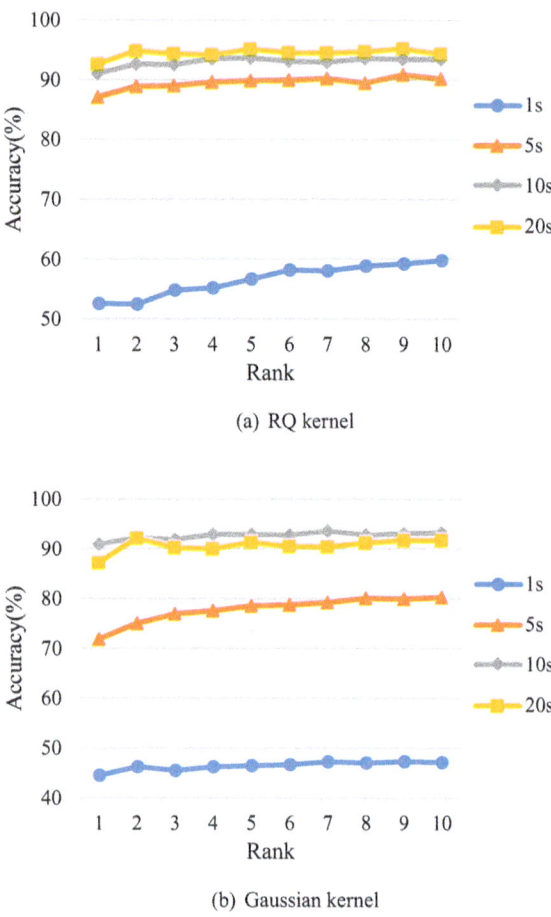

(a) RQ kernel

(b) Gaussian kernel

inherently low-rank. Figure 6.5 presents the results for five subjects on the P300 dataset using the RQ kernel-based GoDec+ algorithm with a rank of 4 and a time length scale of 20 s. A comparison across the four datasets reveals that the time scale has a significant impact on recognition accuracy, particularly in the BCI and P300 datasets. For these datasets, longer time scales result in higher recognition accuracy. This is likely due to the limited number of channels in these datasets, which collect less information. Longer time scales generate larger samples, which contain more information, compensating for the reduced number of channels. Therefore, to ensure high recognition accuracy when using EEG signals collected with a small number of channels, it is essential to extend the acquisition time to obtain a sufficient number of samples.

6.3 Brain Fingerprint Identification with LRMD

Fig. 6.4 Recognition accuracy (%) with RQ kernel on different datasets

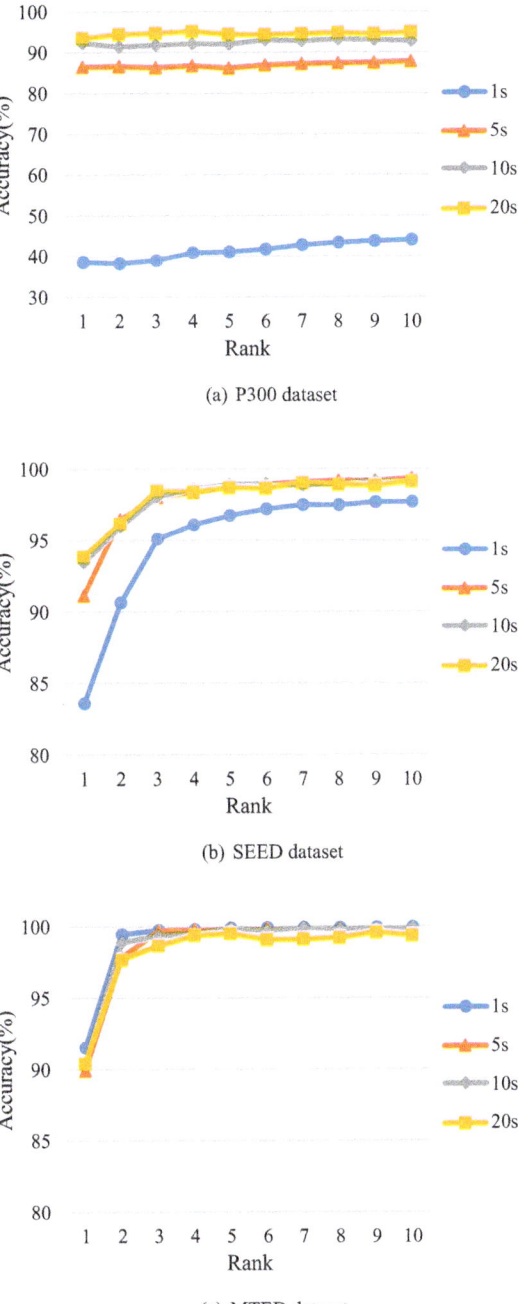

(a) P300 dataset

(b) SEED dataset

(c) MTED dataset

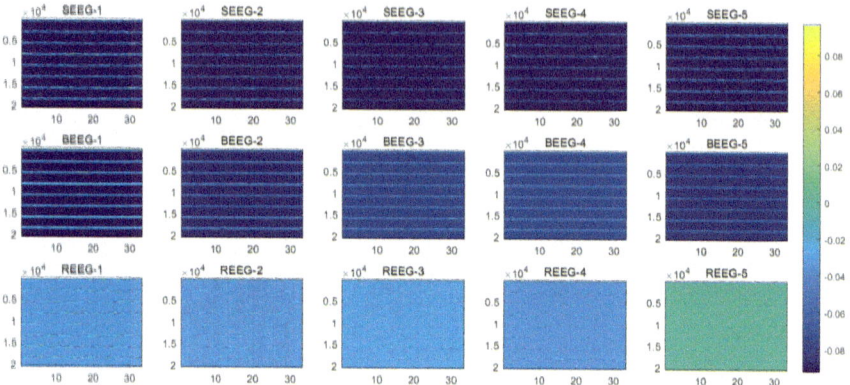

Fig. 6.5 The EEG-spectrum (SEEG), BEEG, and REEG of five subjects are shown using the RQ kernel-based GoDec+ algorithm on the P300 dataset. The results for each subject are displayed in a separate column

6.3.3 Performances of LRMD with Different Kernels

A comparative experiment between the RQ kernel and the Gaussian kernel was conducted using the BCI Graz dataset A. The results are shown in Figs. 6.3, 6.6, and 6.7. As observed in Fig. 6.6, the RQ kernel significantly reduces computation time compared to the Gaussian kernel, achieving similar recognition accuracy. This is particularly evident for lower rank values and longer time scales, indicating that the RQ kernel-based GoDec+ algorithm is more efficient.

The line charts in Fig. 6.7 reveal several trends. First, at a time length scale of 1 s, changes in rank have little impact on time expense, with recognition time remaining steady at around 0.2 s. At a 5-s time scale, the RQ kernel operates at roughly twice

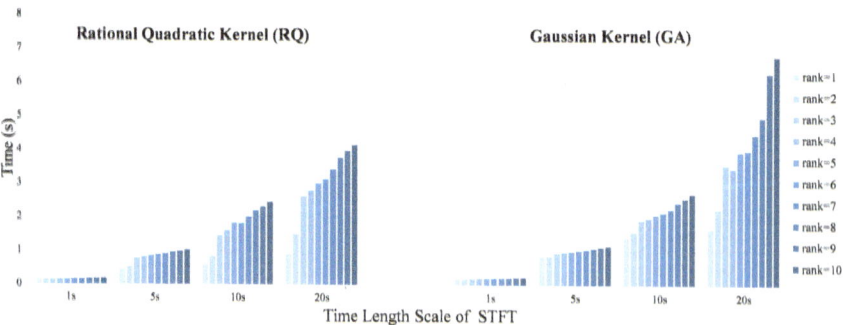

Fig. 6.6 The time comparison between the RQ kernel and the Gaussian kernel on the BCI Graz dataset A is shown in the histogram. The chart illustrates the time required to identify a test sample at specific ranks and time length scales. The left section of the histogram represents the performance of the RQ kernel, while the right section represents the performance of the Gaussian kernel

6.4 Conclusion

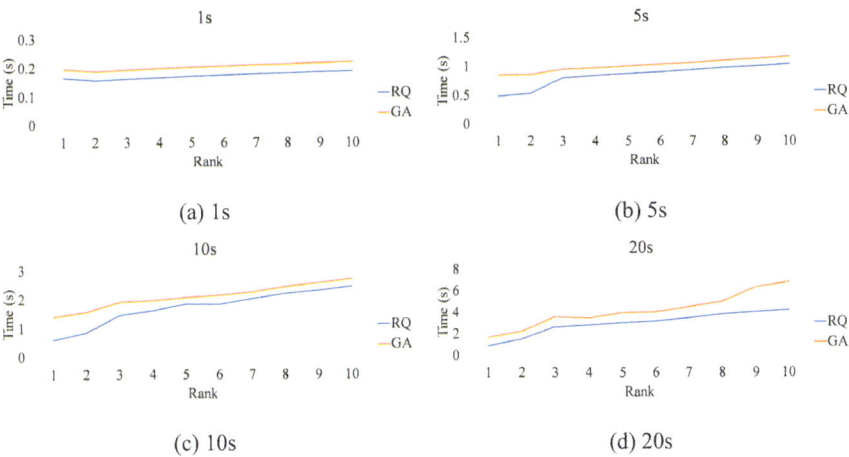

Fig. 6.7 The four line charts show the variation in time spent by the RQ kernel and the Gaussian kernel at different ranks for a specific time length scale

the speed of the Gaussian kernel when the rank is very low. As the time length scale increases to 10 s, the effect of rank on time expense becomes more pronounced, yet the RQ kernel consistently outperforms the Gaussian kernel. At the 20-s scale, rank changes significantly influence time costs; however, the RQ kernel still only requires about 70% of the time taken by the Gaussian kernel, maintaining a stable ratio. Across all shorter time scales, the RQ kernel remains more efficient, and as both the time length and rank grow, the improved LRMD algorithm demonstrates stable performance.

6.4 Conclusion

This chapter presented a task-free brain fingerprint identification model based on an efficient low-rank matrix decomposition (LRMD) algorithm, which addresses the limitations of traditional brain fingerprint identification methods that rely on specific tasks. By integrating low-rank matrix decomposition with sparse representation, the proposed approach enables the successful classification of test data without requiring direct involvement in the training process. This innovative method provides a robust solution for brain fingerprint identification under multiple tasks and emotional states, highlighting its versatility in real-world applications.

The proposed model demonstrates the successful extraction of low-rank, unique brainprints from EEG signals, particularly in scenarios where the data is collected using a small number of channels. Notably, the chapter emphasizes the importance of longer EEG segmentation in such cases to ensure sufficient information for accurate identification. For optimal authentication performance, the method benefits from a careful balance between the rank of the matrix and the time length scale of the data, offering flexibility in practical deployment.

While the approach shows significant promise in terms of accuracy and computational efficiency, it also presents certain limitations. In particular, the trade-off between rank and time length scale requires careful consideration to maintain optimal performance. Moreover, while the method is effective in scenarios with a limited number of channels, longer segmentation times may be necessary to achieve consistently high recognition accuracy.

Overall, the contributions made in this chapter bring us closer to the real-life application of brain fingerprint identification, paving the way for more efficient, task-free identification systems. However, further research is needed to fine-tune the balance between rank and time scale, as well as to explore other potential factors influencing identification accuracy.

References

1. Palaniappan R, Mandic DP (2007) Biometrics from brain electrical activity: a machine learning approach. IEEE Trans Pattern Anal Mach Intell 29(4):738–742
2. Gui Q, Jin Z, Xu W (2015) Exploring EEG-based biometrics for user identification and authentication. In: Signal processing in medicine and biology symposium, pp 1–6
3. Maiorana E, La Rocca D, Campisi P (2016) Eigenbrains and eigentensorbrains: parsimonious bases for EEG biometrics. Neurocomputing 171:638–648
4. Puengdang S, Tuarob S, Sattabongkot T, Sakboonyarat B (2019) EEG-based person authentication method using deep learning with visual stimulation. In: 2019 11th international conference on knowledge and smart technology (KST), pp 6–10, January 2019
5. Bengson JJ, Kelley TA, Zhang X, Wang JL (2014) Spontaneous neural fluctuations predict decisions to attend. J Cogn Neurosci 26(11):2578–2584
6. Gramfort A, Strohmeier D, Haueisen J, Hämäläinen M, Kowalski M (2013) Time-frequency mixed-norm estimates: sparse M/EEG imaging with non-stationary source activations. Neuroimage 70(2):410–422
7. Yang M, Zhang L, Feng X, Zhang D (2011) Fisher discrimination dictionary learning for sparse representation. Proceedings 24(4):543–550
8. Guo K, Liu L, Xu X, Xu D, Tao D (2018) Godec+: fast and robust low-rank matrix decomposition based on maximum correntropy. IEEE Trans Neural Netw Learn Syst 29(6):2323–2336
9. Krystal AD, Prado R, West M (1999) New methods of time series analysis of non-stationary EEG data: eigenstructure decompositions of time varying autoregressions. Clin Neurophysiol 110(12):2197–2206
10. Lu Y, Jiang H, Liu W (2017) Classification of EEG signal by STFT-CNN framework: identification of right-/left-hand motor imagination in BCI systems. In: International conference on computer engineering and networks
11. Daubechies I (2015) The wavelet transform, time-frequency localization and signal analysis. J Renew Sustain Energy 36(5):961–1005
12. Murugappan M, Murugappan S, Balaganapathy, Gerard C (2014) Wireless EEG signals based neuromarketing system using Fast Fourier Transform (FFT). In: IEEE international colloquium on signal processing & ITS applications, pp 25–30
13. Peng Y, Bao-Liang L (2017) Discriminative extreme learning machine with supervised sparsity preserving for image classification. Neurocomputing 261:242–252
14. Bouwmans T, Sobral A, Javed S, Jung SK, Zahzah EH (2016) Decomposition into low-rank plus additive matrices for background/foreground separation. Comput Sci Rev 23(C):1–71

References

15. Ebadi SE, Izquierdo E (2015) Approximated RPCA for fast and efficient recovery of corrupted and linearly correlated images and video frames. In: International conference on systems, signals and image processing, pp 49–52
16. Yuan X, Yang J (2009) Sparse and low-rank matrix decomposition via alternating direction methods. Preprint 12:2
17. Zhou T, Tao D (2011) GoDec: randomized lowrank & sparse matrix decomposition in noisy case. In: International conference on machine learning, ICML 2011, Bellevue, Washington, USA, June 28–July, pp 33–40
18. Nikolova M, Ng MK (2005) Analysis of half-quadratic minimization methods for signal and image recovery. Society for Industrial and Applied Mathematics
19. Zhou T, Tao D (2013) Greedy bilateral sketch, completion & smoothing. JMLR Org, 650–658
20. Chander S, Vijaya P, Dhyani P (2017) Multi kernel and dynamic fractional lion optimization algorithm for data clustering. AEJ—Alexandria Engineering Journal 57(1)
21. Merzbach UC, Boyer CB (2011) A history of mathematics. Wiley
22. Liang J, Wang Y, Zeng X (2015) Robust ellipse fitting via half-quadratic and semidefinite relaxation optimization. IEEE Trans Image Process 24(11):4276–4286
23. Müller-Putz GR, Schlogl A, Pfurtscheller G, Brunner C, Leeb R (2008) BCI competition 2008—Graz data set A
24. Riccio A, Simione L, Schettini F, Pizzimenti A, Inghilleri M, Belardinelli MO, Mattia D, Cincotti F (2013) Attention and p300-based BCI performance in people with amyotrophic lateral sclerosis. Front Hum Neurosci 7(1):732
25. Zheng W-L, Bao-Liang L (2015) Investigating critical frequency bands and channels for EEG-based emotion recognition with deep neural networks. IEEE Trans Auton Mental Dev 7(3):162–175

Open Access This chapter is licensed under the terms of the Creative Commons Attribution-NonCommercial-NoDerivatives 4.0 International License (http://creativecommons.org/licenses/by-nc-nd/4.0/), which permits any noncommercial use, sharing, distribution and reproduction in any medium or format, as long as you give appropriate credit to the original author(s) and the source, provide a link to the Creative Commons license and indicate if you modified the licensed material. You do not have permission under this license to share adapted material derived from this chapter or parts of it.

The images or other third party material in this chapter are included in the chapter's Creative Commons license, unless indicated otherwise in a credit line to the material. If material is not included in the chapter's Creative Commons license and your intended use is not permitted by statutory regulation or exceeds the permitted use, you will need to obtain permission directly from the copyright holder.

Chapter 7
Multi-task and Single-Session Recognition with Residual Multi-scale Neural Network

Abstract This chapter presents an advanced brain fingerprint identification method that leverages electroencephalography (EEG) signals, referred to as the Residual and Multi-scale Spatio-Temporal Convolution Neural Network (RAMST-CNN). Brain fingerprinting, a unique approach to biometric identification, relies on the inherent characteristics of EEG signals to identify individuals based on their cognitive and neurological patterns. The RAMST-CNN model integrates several cutting-edge deep learning techniques to enhance its performance in feature extraction, including Residual Learning (RL), Multi-scale Grouping Convolution (MGC), Global Average Pooling (GAP), and Batch Normalization (BN). These components work synergistically to enable robust extraction of spatio-temporal features from EEG data, making the model highly effective in capturing both spatial and temporal dynamics that are crucial for brain fingerprinting. The task-independent design of the RAMST-CNN significantly alleviates the complexities traditionally associated with manual feature selection and extraction, which have often been a bottleneck in previous brain fingerprinting systems. By automating feature learning, the method minimizes human intervention and ensures consistent feature representation across different cognitive tasks. The model's efficiency is further enhanced by its lightweight architecture, which facilitates high-performance recognition without the need for large computational resources. Extensive comparative evaluations across a range of datasets and against various state-of-the-art methods demonstrate the superior performance of RAMST-CNN in terms of accuracy, robustness, and generalization. The results underscore the model's potential for real-world brain fingerprinting applications, highlighting its ability to deliver high recognition accuracy in both single-session and multi-task scenarios.

7.1 Introduction

Personal identification technology plays a vital role in domains such as counter-terrorism, riot control, and the safeguarding of personal information, particularly in the era of rapid Internet development. Traditional personal identification methods, which rely on physical tokens like ID cards and passwords, are inherently susceptible to risks such as loss, theft, or unauthorized use. In response to these challenges, modern personal identification systems increasingly leverage biometrics—physiological traits such as facial recognition, fingerprints, palm prints, iris patterns, and even sweat. Despite their widespread adoption, these biometric methods remain vulnerable to forgery or replication, raising significant concerns over their reliability and security. Electroencephalography (EEG) signals, which represent the dynamic electrophysiological activity of the brain, reflect the collective electrical activity of neurons recorded from the scalp or cerebral cortex [1]. Compared to traditional biometric modalities, EEG-based brain fingerprint identification offers several advantages, including high specificity, resilience to counterfeiting, robustness against stress-induced variations, and non-replicability. Moreover, EEG-based methods are exclusively applicable to in-vivo detection, making them inherently secure and a compelling area of research [2–4].

Current approaches to brain fingerprint identification based on specific tasks can be broadly categorized into four methods: resting state [5–9], motor imagery [2], visual-evoked potentials [4, 10, 11], and event-related potentials. While these methods demonstrate effectiveness, their inherent task dependency represents a critical limitation.

The task-dependent nature of these methods stems from two primary factors. First, subjects are required to perform specific tasks under controlled conditions. For example, resting-state brain fingerprinting necessitates participants to remain calm in a quiet, relaxed environment, whereas evoked EEG methods demand active engagement during data collection, such as watching videos or listening to audio stimuli. These requirements can induce fatigue or boredom and are often unsuitable for individuals with cognitive impairments.

Second, the brainprint features used for identification must be carefully tailored to the specific task and data collection scenario. Features such as power spectral density (PSD), autoregressive coefficients [12], sparse representations [13], and wavelet coefficients are commonly extracted. The choice of features often depends on the task context. For instance, PSDs in the beta band (12–30 Hz) are typically extracted when mood variations are anticipated, while gamma band (30–50 Hz) features are associated with multitasking scenarios, further illustrating task-dependent feature extraction. To overcome these limitations, achieving task independence in data processing is critical, and automatic feature extraction methods provide a promising solution.

Convolutional Neural Networks (CNNs), as powerful feature learning algorithms, can automatically extract discriminative features from data, thereby eliminating the need for complex feature engineering. CNNs have gained significant traction in brain fingerprint identification [14, 15]. However, traditional CNN architectures often involve a large number of parameters, requiring extensive training data and prolonged training times. These constraints limit their practicality in real-world applications, underscoring the need for more efficient models tailored to brain fingerprinting tasks.

In this chapter, a novel multi-task brain fingerprint identification method is designed based on a lightweight Convolutional Neural Network (CNN). The feature extraction capability of the network is significantly enhanced through the integration of Residual Learning (RL) [16] and Multi-Scale Grouping Convolution (MGC) [17]. Additionally, the adoption of Global Average Pooling (GAP) [18] as a substitute for traditional fully connected layers substantially reduces the number of parameters, effectively mitigating the risk of overfitting. Batch Normalization (BN) [19] is further employed to accelerate network training and improve stability. Collectively, these innovations contribute to the development of a robust and efficient framework for EEG-based personal identification, offering valuable insights for real-world applications. The primary contributions of this chapter are as follows:

- Elimination of complex, task-dependent feature engineering: By removing the dependency on specific data collection conditions, the proposed method overcomes the limitations of traditional brain fingerprint identification approaches.
- Development of a lightweight and efficient network architecture: The method combines advanced techniques to construct a lightweight network for brain fingerprint identification while introducing automated data processing methods, thus avoiding the subjectivity and complexity associated with manual selection and extraction of EEG brainprint features.
- Demonstration of high accuracy and efficiency: The proposed approach achieves superior recognition accuracy and rapid processing times across various datasets, showcasing its practical utility and effectiveness.

7.2 Residual and Multi-scale Spatio-Temporal Convolution Neural Network

RL and MGC are highly effective techniques for enhancing the performance of convolutional layers, particularly in extracting spatio-temporal features. GAP serves as an excellent alternative to fully connected layers, helping to mitigate overfitting. BN further contributes by accelerating and stabilizing the training process. By integrating RL, MGC, GAP, and BN, the Residual and Multi-scale Spatio-Temporal Convolution Neural Network (RAMST-CNN) achieves superior performance in feature extraction and model optimization.

7.2.1 Residual Learning

The basic residual unit, illustrated in Fig. 7.1, embodies the concept of Residual Learning (RL). In this structure, the input is denoted as x, while $F(x)$ represents the original output computed by the residual unit. After calculating $F(x)$, the input x is

Fig. 7.1 Basic residual unit

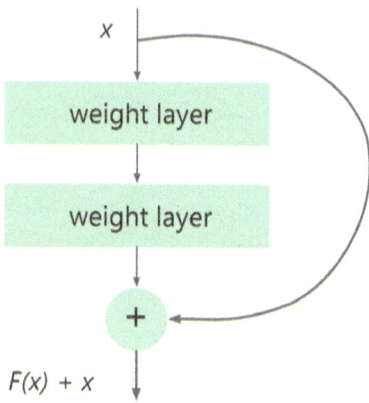

directly added to $F(x)$, producing the final output of the network. This process, often referred to as a constant transformation, ensures that the resulting $F(x)$ captures the residual between the actual output and the input.

The constant transformation effectively mitigates the challenges associated with deeper network architectures. By preserving the identity mapping, the depth of the network becomes virtually transparent to the model, allowing deeper networks to maintain at least the same representational capacity as shallower ones. Simultaneously, the network parameters are continuously updated, enabling the model to learn and enhance its representational power.

In the RAMST-CNN architecture, which comprises multiple layers, the application of RL plays a critical role in reducing the risk of gradient vanishing. This ensures that the network remains capable of learning and optimizing even with increased depth, enhancing its overall performance in extracting spatio-temporal features.

7.2.2 Multi-scale Grouping Convolution

In this section, multi-electrode EEG signals are utilized, which inherently contain rich spatio-temporal information. Convolution kernels of varying sizes are employed to capture diverse spatio-temporal features from the two-dimensional EEG signal inputs. The structure of the Multi-Scale Grouping Convolution (MGC) unit is illustrated in Fig. 7.2.

The MGC unit processes the input through convolution operations using kernels of multiple sizes, such as 5×5, 3×3, 1×1. The outputs from these convolutions are then concatenated along the channel dimension to integrate features across multiple scales. Additionally, the output from the first convolutional layer is included in this concatenation process. This inclusion is critical because the features extracted at different hierarchical levels of the network vary significantly. Typically, lower-level features are more representative of raw spatial information, whereas higher-level

7.2 Residual and Multi-scale Spatio-Temporal Convolution Neural Network

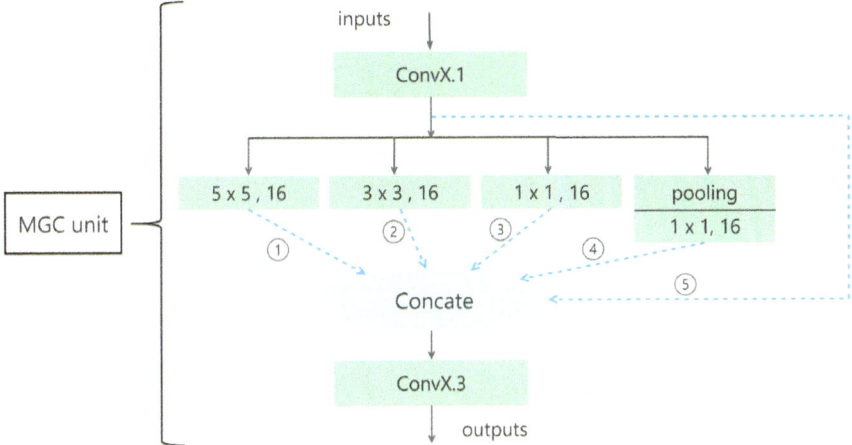

Fig. 7.2 Kernels of multi-scale grouping convolution. ①②③④⑤ are merged in the channel dimension, where 5×5, 3×3, 1×1 represent the size of the convolution kernels, and the number of 16 represents the number of convolution kernels, which is the same as the number of output channels

features become increasingly abstract and task-specific as they pass through deeper layers of the network.

For instance, in applications such as image style transfer, lower-level features are often leveraged for content reconstruction to preserve structural integrity, while higher-level features undergo nonlinear transformations, potentially leading to distortions [20]. In the MGC unit, the final convolutional layer synthesizes these multi-scale features and reduces their dimensionality, achieving a balance between feature richness and computational efficiency.

7.2.3 Global Average Pooling

Pooling operations are commonly used in neural networks to downsample feature maps, thereby reducing their spatial dimensions and computational complexity. For a pooling kernel with size k in a particular dimension, the output scale in the corresponding dimension is approximately $1/k$ of the input. Global Pooling (GP) differs from traditional pooling layers primarily in the scale of the pooling kernel. In GP, the kernel size matches the spatial dimensions of the entire feature tensor, excluding the channel dimension.

During GP, convolution operations are performed between the feature maps of the input tensor and corresponding pooling windows defined by the GP kernel. The output is a vector whose length equals the number of convolution kernels in the preceding layer. GP can be categorized into Global Maximum Pooling (GMP) and Global

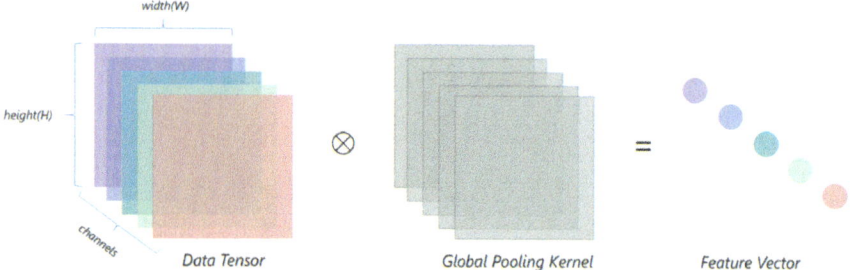

Fig. 7.3 Diagram of a global pooling layer. Each feature map belonging to the input tensor is pooled with the same scale pooling kernel, becoming a value, and the data tensor becomes consequently a vector

Average Pooling (GAP). In this work, GAP is employed due to its ability to flatten input feature tensors into vectors, providing a compact representation of the original data. This approach eliminates the need for fully connected layers, significantly reducing the risk of overfitting.

The GAP operation involves aggregating values within each feature map using the formula:

$$O_k = \frac{1}{H \times Q} \sum_i^H \sum_j^W F_{ij}, \tag{7.1}$$

where O_k represents the output value for the k-th channel, H and W denote the height and width of the feature map, respectively, and F_{ij} represents the value at the i-th row and the j-th column of the feature map. The process of Global Pooling is illustrated in Fig. 7.3.

7.2.4 Batch Normalization

Batch Normalization (BN) is a widely utilized algorithm in DL research, primarily designed to address the issue of internal covariate shift. Internal covariate shift refers to the problem where the distribution of network inputs changes during training, causing difficulties in feature learning and slowing down the training process.

BN normalizes the input data for each mini-batch, ensuring that the data distribution is adjusted in a manner that facilitates faster convergence. During the training phase, BN introduces two learnable hyperparameters: the scale factor and the shift parameter. These parameters allow the network to adapt the distribution of the input data to be more conducive to the activation function's sensitive region, effectively minimizing the risk of vanishing gradients. This self-adaptation of input data distribution leads to improved model performance and often results in faster convergence.

In the context of this study, BN is incorporated into all convolutional layers of RAMST-CNN. Specifically, it is applied before the activation function, ensuring that the activation function can more effectively update the network weights during the training process.

7.2.5 Structure of Network

The overall architecture of the network is designed by integrating key concepts from RL, MGC, GAP, and BN. The network structure is illustrated in Fig. 7.4. Initially, the input is processed by a standard convolutional layer, followed by the MGC unit to extract diverse spatio-temporal features. These features are then fused along the channel dimension, facilitating feature integration. Afterward, the data passes through another conventional convolutional layer. The residual learning mechanism is applied by directly adding the input to the output of the convolutional kernel, which promotes the learning of residual mappings and aids in the network's convergence. Next, GAP is employed, functioning similarly to a flattening layer, and the final layer consists of a fully connected layer that adjusts the network output to match the number of classes. ReLU is used as the activation function across all layers to introduce non-linearity. The entire network is implemented using the TensorFlow framework, and the training process is accelerated using an NVIDIA GTX 1080 Ti graphics processing unit (GPU).

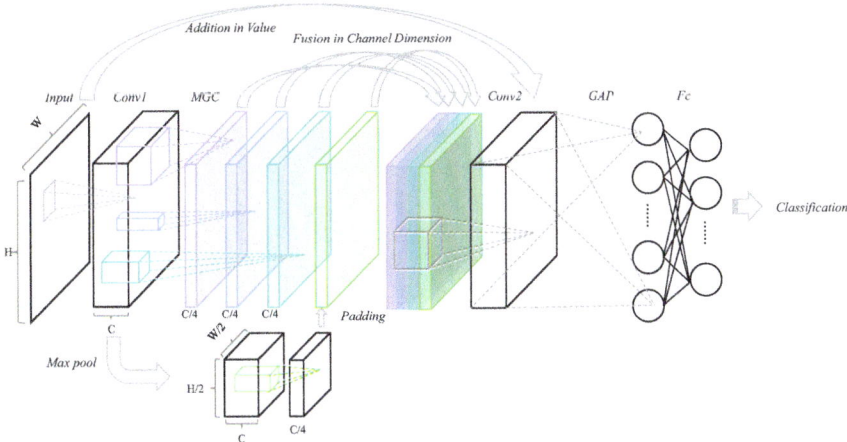

Fig. 7.4 Network structure of RAMST-CNN. The inputs are two-dimensional EEG signals. H, W, and C are the height, width, and number of channels of the feature tensor, respectively. In our method for brain fingerprint identification, C has a default value of 64. *Conv1* represents a convolution operation. The meaning of *conv2* is similar, etc. *Addition in value* implements RL. *Fusion in Channel Dimension* and *Padding* implement MGC; *Fc* denotes a fully connected layer

7.2.6 Training Settings

The specific training parameters are detailed in Table 7.1. The number of training epochs was determined based on an analysis of the loss curve derived from the validation dataset. The initial learning rate was set to 0.001, a standard value widely used in deep learning models. The batch size was chosen from a range of values, including 8, 16, 32, 64, 128, and 256. A very small batch size was found to result in unstable training due to its inability to capture sufficient data to represent the entire dataset, leading to inconsistent gradient descent directions across batches. Conversely, a larger batch size reduced the frequency of model updates, diminishing overall training efficiency and increasing the risk of memory overflow. Based on these considerations, an empirical batch size of 128 was selected as the optimal compromise. Furthermore, the Adam optimizer was employed, taking advantage of its adaptive learning rate capabilities to improve the convergence of the model during training.

7.3 Brain Fingerprint Identification with RAMST-CNN

7.3.1 Description of Datasets

A total of five datasets were utilized, including the BCI Graz dataset A (BCI) [21], the P300 EEG Dataset in 2013 (P300) [22], the SJTU Emotion EEG Dataset (SEED) [23], the multi-task EEG dataset (MTED), which was collected by our laboratory [24], and a dataset for emotion analysis using EEG, physiological and video signals (DEAP) [25], a widely used dataset in emotion recognition research. These datasets include data collected across various tasks, such as resting state, motor imagery, visual-evoked potential tasks, and event-related potential experiments.

A summary of the datasets is provided in Table 7.2.

Table 7.1 Parameters of training

Training parameters	Values
Training epochs	50
Learning rate	0.001
Batch size	128
Optimizer	Adam

7.3 Brain Fingerprint Identification with RAMST-CNN

Table 7.2 Brief introduction of datasets involved

Dataset	Task	Participants	Electrodes	SF[f] (Hz)
BCI	MI[a]	9	22	250
P300	ER[b]	8	8	256
SEED	VE[c]	15	62	1000
MTED	MT[d]	15	62	1000
DEAP	VAE[e]	32	32	128

[a] MI: Movement imaging, [b] ER: Event related, [c] VE: Visual evoked, [d] MT: Multiple tasks, [e] VAE: Visual and audio evoked, [f] SF: Sampling frequency

7.3.2 Data Pre-processing

The preprocessing begins with segmenting each dataset based on the product of the number of electrodes and the sampling frequency. This segmentation is followed by downsampling to a predefined height (H) and width (W). In this context, the height corresponds to the number of electrodes, while the width is determined by downsampling the signal to half or less of the original sampling frequency. In this study, downsampling rates of 0.5 and 0.25 were applied, with the width corresponding to the length of a 1-second signal after downsampling in the time domain. Specifically, a downsampling rate of 2 is applied to signals with a sampling frequency of 128 Hz, and a rate of 4 is used for signals sampled at 200 Hz or 250 Hz. This reduction in sampling rate helps simplify the size of each sample and facilitates computational efficiency.

After downsampling, unique labels are assigned to each individual to create a complete sample set. This set is then randomly divided into training and testing sets, each comprising 50% of the total samples. This 50/50 split is an intentional departure from the more commonly used 80/20 or 90/10 training-to-testing splits typically employed in deep learning tasks. The rationale for this decision lies in the lightweight nature of the network, which will be discussed further in the subsequent sections. With fewer parameters in the network, the model is less prone to overfitting, even with a smaller training set.

Each EEG signal is first sliced into small fragments in the time domain and then each fragment representing a sample is flattened into a vector. Next, all samples are labeled with different labels according to the identity of the subject, and an entire data set is formed. Finally, shuffling is adopted before all samples are divided into training and testing sets. Each multi-electrode EEG is treated as a 2-dimensional image, encapsulating spatio-temporal information.

During training, one-tenth of the training data is randomly selected from the training set to form a validation set. After every ten batch updates, the network is evaluated on the validation set to assess whether the model is learning correctly. It is important to note that the validation set is not involved in the training process itself. To standardize the data, each input sample is mean-centered and normalized by dividing by the standard deviation, a technique commonly referred to as Z-score

Table 7.3 Size of a single sample and the division of the dataset

Dataset	Height (H)	Width (W)	Training sample	Validation sample	Testing sample
BCI	22	125	3499	389	3888
P300	8	128	4889	543	5432
SEED	62	100	4556	506	5062
MTED	62	100	5400	600	6000
DEAP	32	128	13608	1512	15120

standardization. This method is widely used to transform data with different magnitudes into a comparable scale, making it easier to analyze. The calculations for this standardization process are as follows:

$$\mu = \frac{1}{H \times W} \sum_i^H \sum_j^W x_{ij}, \tag{7.2}$$

$$s = \sqrt{\frac{\sum_i^H \sum_j^W (x_{ij} - \mu)^2}{H \times W - 1}}, \tag{7.3}$$

$$y_{ij} = \frac{x_{ij} - \mu}{s}, \tag{7.4}$$

where x_{ij} represents the signal value at the i-th row and j-th column of a sample, and μ and s denote the mean value and standard deviation value, respectively. The size of the sample and the division of the datasets are detailed in Table 7.3.

7.3.3 Performances of Multi-task and Single-Session Brain Fingerprint Identification

To ensure the reliability of the results, both the training and testing datasets were constructed ten times, and the experiments were repeated with the average results calculated to assess the robustness of the method.

The results, summarized in Table 7.4, demonstrate that very high and stable recognition accuracy can be achieved across all datasets, highlighting the effectiveness and consistency of our approach. Despite the relatively small number of participants in the BCI and P300 datasets, the brain fingerprint identification accuracy on the other datasets, particularly DEAP (with 32 participants), indicates that the number of subjects does not significantly affect performance.

7.3 Brain Fingerprint Identification with RAMST-CNN

Table 7.4 Test accuracy (%) on each dataset

Dataset	Accuracy (%)
MTED	100.00 ± 0.00
SEED	99.78 ± 0.07
P300	99.33 ± 0.04
BCI	99.68 ± 0.06
DEAP	99.94 ± 0.03

Table 7.5 Brain fingerprint identification accuracy of each method on BCI, P300, and MTED datasets

Dataset	Methods	Average accuracy (%)
BCI	LRMD [24]	95.45
	AR/BP neural network [12]	81.20, 82.10, 82.80, 90.60
	RAMST-CNN	**99.68**
P300	LRMD [24]	95.34
	RAMST-CNN	**99.33**
MTED	LRMD [24]	99.77
	RAMST-CNN	**100.00**

The methods used for EEG-based brain fingerprint identification are compared to those used in other studies. Kong et al. [24] employed low-rank matrix decomposition (LRMD) to extract the background EEG (BEEG) with low-rank characteristics, constructing a subspace from all samples. The optimal reconstruction coefficients for the test sample were then determined using the maximum cross-entropy criterion (MGCC), which uses the cross-entropy between test and reconstructed samples for classification. The accuracy achieved was 95.45% on the BCI dataset and 95.34% on the P300 dataset. For the MTED dataset, with a matrix rank of 8, the highest accuracy reached 99.98%, with an average accuracy of 99.76%. Hu et al. [12] divided the BCI motion imagery dataset into four subsets corresponding to different imagined tasks (left hand, right hand, leg, and tongue). They extracted features from the motor imagery EEG signals using the regression coefficient, energy spectral density, phase synchronization, and linear complexity. A backpropagation (BP) neural network was then used for classification, achieving accuracies of 81.2, 82.1, 82.8, and 90.6% on each subset. Table 7.5 presents the recognition accuracy of each method on the BCI, P300, and MTED datasets.

As shown in Table 7.5, the proposed method, based on RAMST-CNN, outperforms the other methods in terms of both recognition accuracy and efficiency. Although the LRMD method also achieves task independence and performs well in brain fingerprint identification, it requires the rank of the low-rank matrix to be predetermined. Furthermore, achieving higher accuracy necessitates longer sample durations, which poses challenges for the design of a real-time EEG-based brain fingerprint identification system. In contrast, the RAMST-CNN-based approach provides a more efficient and flexible solution.

The proposed method is compared with several state-of-the-art techniques used in other studies, specifically based on the DEAP dataset. Zhang et al. [26] introduced a method using an attention-based recurrent neural network (ARNN), where long short-term memory (LSTM) was employed to implement the memory mechanism. In this approach, one-dimensional EEG signals, after being filtered into the δ subband, serve as inputs to the network. Mao et al. [27] used a CNN, which primarily consisted of two convolutional layers followed by two fully connected layers. Two-dimensional EEG signals, each with a time window of 1 s, were used as inputs in their approach. The various methods and their corresponding results are shown in Fig. 7.5. It is important to note that the results presented are the average outcomes of ten independent experiments.

The proposed method achieved the highest identification accuracy, reaching 99.94 $\pm$ 0.03%, outperforming both the ARNN and CNN methods, which achieved 97.41 $\pm$ 1.90% and 96.01 $\pm$ 0.09%, respectively. Additionally, the results from the ARNN method, which used one-dimensional input vectors, exhibited instability. In contrast, our model, which extracts spatio-temporal features from the data, retains more relevant information about each subject. Compared to the CNN approach proposed by Mao et al. [27], the RAMST-CNN demonstrated superior feature extraction capabilities, leading to higher and more stable performance.

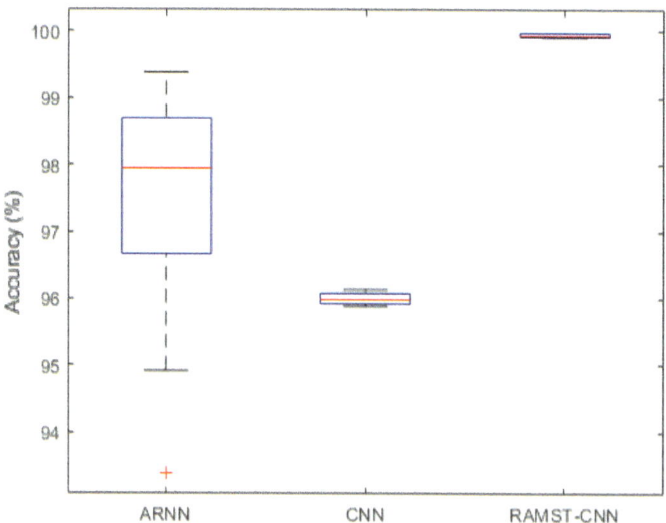

Fig. 7.5 Comparison on DEAP. The five key values of one box in this figure (from top to bottom), the largest value (top line), upper quartile (upper edge of the box), median (middle red line), lower quartile (lower edge of the box), and smallest value (bottom line) except for the outlier (red cross)

7.3.4 Effect of Each Component

BN is a widely used and powerful technique in DL, which is why it serves as the default setting for the network. However, the contributions of Residual Learning (RL), Multi-scale Grouping Convolution (MGC), and Global Average Pooling (GAP)—as structural components of the network—are not inherently guaranteed to improve performance. Therefore, an ablation study was conducted to assess the individual impact of these components. The experiments were carried out under the following conditions: (1) proposed model (RL + MGC + GAP); (2) without RL (MGC + GAP); (3) without MGC (RL + GAP); (4) without GAP (RL + MGC); (5) baseline (without RL, MGC, and GAP). To simplify the study, the ablation tests were performed solely on the P300 dataset, which is internationally recognized and commonly used for such evaluations. The training conditions for this experiment are summarized in Table 7.1.

The results of the ablation study on the P300 dataset are presented in Fig. 7.6. The proposed model (RL + MGC + GAP) achieved the highest and most stable performance, with an accuracy of $99.33 \pm 0.04\%$. The comparison across the different model configurations highlights the importance of RL, MGC, and GAP in improving model performance. Notably, removing GAP resulted in the largest accuracy drop. When compared to the baseline model, which lacks all three components (RL, MGC, and GAP), the enhanced identification accuracies validate the effectiveness of these techniques.

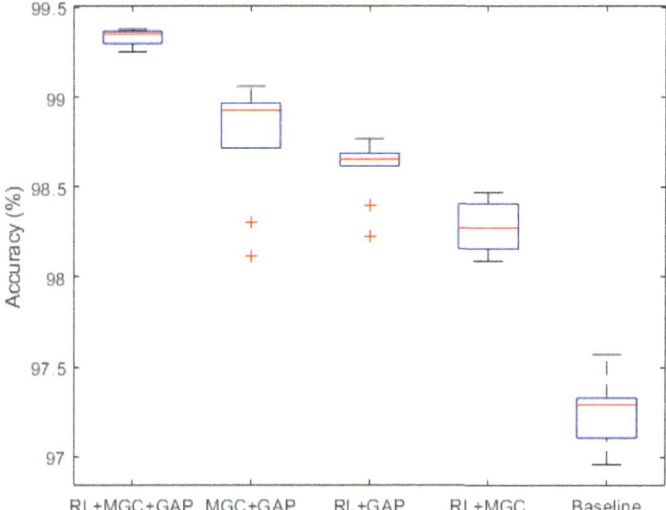

Fig. 7.6 Comparison of the results of the ablation study on P300. Note that the lower whiskers of "RL + GAP" and "RL + MGC" disappeared, which can be attributed to the two outliers in these two groups (ten values in total in each group), represented by small red pluses in the figure

7.4 Conclusion

In this chapter, we have introduced an innovative approach to brain fingerprint identification using multi-electrode EEG signals. By leveraging the Residual and Multi-scale Spatio-Temporal Convolution Neural Network (RAMST-CNN), which integrates advanced deep learning techniques such as Multi-scale Grouping Convolution (MGC), Residual Learning (RL), Global Average Pooling (GAP), and Batch Normalization (BN), we have demonstrated a method capable of automatically extracting and fusing spatio-temporal features at multiple scales. This end-to-end solution eliminates the need for manual feature extraction, thereby enhancing task independence during the data processing phase. Compared to traditional brain fingerprinting methods, RAMST-CNN offers superior performance, particularly in terms of its lightweight architecture, which is conducive to real-world applications where computational efficiency is paramount.

Despite these significant advantages, there are inherent limitations that must be addressed for the method's broader applicability. One key challenge lies in the size and diversity of the training datasets. Current datasets often suffer from a limited number of participants, which restricts the generalization capability of the model. As the pool of participants grows, the network will need to handle an increasingly larger set of features, potentially necessitating further refinements in the architecture. Additionally, although the RAMST-CNN model is designed to be lightweight, the scale of the training data still presents challenges in terms of computational resources. To effectively train the model with smaller datasets while maintaining accuracy, future work will focus on optimizing the network architecture and exploring new experimental strategies.

Furthermore, while the RAMST-CNN excels in task-independent feature extraction, future developments should explore its adaptability across diverse cognitive tasks and real-world scenarios. As EEG-based brain fingerprinting systems continue to evolve, overcoming the limitations related to dataset size and ensuring robust performance across various conditions will be critical for the widespread adoption of this technology.

References

1. Paranjape RB, Mahovsky J, Benedicenti L, Koles' Z (2001) The electroencephalogram as a biometric. In: Proceedings of the Canadian conference on electrical and computer engineering on Toronto, Canada, vol 2, pp 1363–1366
2. Marcel S, Millan J (2007) Person authentication using brainwaves (EEG) and maximum a posteriori model adaptation. IEEE Trans Pattern Anal Mach Intell 29(4):743–752
3. Nevado-Holgado AJ, Marten F, Richardson MP, Terry JR (2012) Characterising the dynamics of EEG waveforms as the path through parameter space of a neural mass model: application to epilepsy seizure evolution. Neuroimage 59(3):2374–2392
4. Yeom SK, Suk HI, Lee SW (2013) Person authentication from neural activity of face-specific visual self-representation. Pattern Recognit 46(4):1159–1169

References

5. Jian Y, Shu-Shen C, Hao-Ran H, Pei-Peng L, Ning Z (2015) Dynamic functional connectivity of electroencephalogram in the resting state. Acta Physica Sinica 64(5):58701
6. Poulos M, Rangoussi M, Alexandris N (1999) Neural network based person identification using EEG features. In: Proceedings of the IEEE international conference on acoustics, speech, and signal processing. Proceedings. (ICASSP99) on Phoenix, AZ, USA, vol 2, pp 1117–1120
7. Liu Q-Y (2013) A study on Resting Electroencephalogram (EEG) based biometrics in ubiquitous environment. PhD thesis, Lanzhou University
8. Poulos M, Rangoussi M, Chrissikopoulos V, Evangelou A (1999) Person identification based on parametric processing of the EEG. In: Proceedings of the ICECS'99. 6th IEEE international conference on electronics, circuits and systems on Pafos, Cyprus, vol 1, pp 283–286
9. Maiorana E, La Rocca D, Campisi P (2016) Eigenbrains and eigentensorbrains: parsimonious bases for EEG biometrics. Neurocomputing 171:638–648
10. Das K, Zhang S, Giesbrecht B, Eckstein MP (2009) Using rapid visually evoked EEG activity for person identification. In: Proceedings of the annual international conference of the IEEE engineering in medicine and biology society on minneapolis, MN, USA, pp 2490–2493
11. Gui Q, Jin Z, Xu W (2014) Exploring EEG-based biometrics for user identification and authentication. In: Proceedings of the IEEE signal processing in medicine and biology symposium (SPMB) on Philadelphia, PA, USA, pp 1–6
12. Jian-feng H, Xue-cai B (2012) Person identification based on elecroencephalogram signals. J Clin Rehabil Tissue Eng Res 13(17):62–66
13. Peng Y, Bao-Liang L (2017) Discriminative extreme learning machine with supervised sparsity preserving for image classification. Neurocomputing 261:242–252
14. Chen JX, Mao ZJ, Yao WX, Huang YF (2019) EEG-based biometric identification with convolutional neural network. Multimed Tools Appl 79(331):1–21
15. Das BB, Kumar P, Kar D, Ram SK, Babu KS, Mohapatra RK (2019) A spatio-temporal model for EEG-based person identification. Multimed Tools Appl 78(19):28157–28177
16. He K, Zhang X, Ren S, Sun J (2016) Deep residual learning for image recognition. In: Proceedings of the IEEE conference on computer vision and pattern recognition on Las Vegas, NV, USA, pp. 770–778
17. Szegedy C, Liu W, Jia Y, Sermanet P, Reed S, Anguelov D, Erhan D, Vanhoucke V, Rabinovich A (2015) Going deeper with convolutions. In: Proceedings of the IEEE conference on computer vision and pattern recognition on Boston, MA, USA, pp 1–9
18. Lin M, Chen Q, Yan S (2013) Network in network. arXiv:1312.4400
19. Ioffe S, Szegedy C (2015) Batch normalization: accelerating deep network training by reducing internal covariate shift. In: Proceedings of the the 32nd international conference on international conference on machine learning in Lille, France, vol 37, pp 448–456
20. Gatys LA, Ecker AS, Bethge M (2016) Image style transfer using convolutional neural networks. In: Proceedings of the IEEE conference on computer vision and pattern recognition on Las Vegas, NV, USA, pp 2414–2423
21. Brunner C, Leeb R, Müller-Putz G, Schlögl A, Pfurtscheller G (2008) BCI Competition 2008– Graz data set A. Institute for knowledge discovery (Laboratory of brain-computer interfaces), Graz University of Technology, vol 16
22. Riccio A, Simione L, Schettini F, Pizzimenti A, Inghilleri M, Belardinelli MO, Mattia D, Cincotti F (2013) Attention and P300-based BCI performance in people with amyotrophic lateral sclerosis. Front Hum Neurosci 7:732
23. Zheng W-L, Bao-Liang L (2015) Investigating critical frequency bands and channels for EEG-based emotion recognition with deep neural networks. IEEE Trans Auton Ment Dev 7(3):162–175
24. Kong X, Kong W, Fan Q, Zhao Q, Cichocki A (2018) Task-independent EEG identification via low-rank matrix decomposition. In: Proceedings of the IEEE international conference on bioinformatics and biomedicine (BIBM) on Madrid, Spain, pp 412–419
25. Koelstra S, Muhl C, Soleymani M, Lee J-S, Yazdani A, Ebrahimi T, Pun T, Nijholt A, Patras I (2011) DEAP: a database for emotion analysis; using physiological signals. IEEE Trans Affect Comput 3(1):18–31

26. Zhang X, Yao L, Kanhere SS, Liu Y, Gu T, Chen K (2018) MindID: person identification from brain waves through attention-based recurrent neural network. Proc ACM Interact, Mob, Wearable Ubiquitous Technol 2(3):1–23
27. Mao Z, Yao WX, Huang Y (2017) EEG-based biometric identification with deep learning. In: Proceedings of the IEEE/EMBS international conference on neural engineering on Shanghai, China, pp 609–612

Open Access This chapter is licensed under the terms of the Creative Commons Attribution-NonCommercial-NoDerivatives 4.0 International License (http://creativecommons.org/licenses/by-nc-nd/4.0/), which permits any noncommercial use, sharing, distribution and reproduction in any medium or format, as long as you give appropriate credit to the original author(s) and the source, provide a link to the Creative Commons license and indicate if you modified the licensed material. You do not have permission under this license to share adapted material derived from this chapter or parts of it.

The images or other third party material in this chapter are included in the chapter's Creative Commons license, unless indicated otherwise in a credit line to the material. If material is not included in the chapter's Creative Commons license and your intended use is not permitted by statutory regulation or exceeds the permitted use, you will need to obtain permission directly from the copyright holder.

Chapter 8
Multi-task and Single-Session with Convolutional Tensor-Train Neural Network

Abstract Deep learning has demonstrated a remarkable ability to extract high-level features and uncover complex latent dependencies, making it highly effective for various tasks. However, the success of deep learning models typically hinges on the availability of large datasets for training, which poses a significant challenge in real-world applications. In the domain of brain fingerprint identification, where data from multiple individuals is often sparse and each class contains only a few samples, deep learning models face difficulties in achieving reliable performance. This chapter presents a Convolutional Tensor-Train Neural Network (CTNN) designed to tackle these challenges in the context of multi-task brain fingerprint identification with limited training samples. The method integrates a convolutional neural network (CNN) with a depthwise separable convolution mechanism to extract local temporal and spatial features from the brainprint, focusing on subtle neural patterns that distinguish individuals. To further enhance the model's ability to capture complex interdependencies, the TensorNet (TN) component is introduced, employing low-rank tensor decomposition to model multilinear interactions between different features. This representation enables the model to efficiently integrate local information into a global feature space with minimal parameters, making it particularly effective in scenarios with small-sample sizes. The CTNN approach not only addresses the challenge of limited data but also excels in multi-task learning, enabling it to extract shared features across different recognition tasks. This capability is crucial for real-world applications where brainprint identification must function across diverse individuals and conditions. Furthermore, the model provides interpretability by identifying key brain regions, with seven specific channels being dominant in the recognition tasks, thus offering valuable insights into the neural biomarkers underlying brain fingerprint identification.

This chapter includes content from "CTNN: A convolutional tensor train neural network for multi-task brainprint recognition" by Xuanyu Jin, et al., available under CC BY-NC-ND 4.0. The original work can be found at https://ieeexplore.ieee.org/abstract/document/9248036. The content is slightly modified.

8.1 Introduction

Brain fingerprint identification can be broadly classified into three categories based on the type of task stimulus presented to the subjects: resting potentials (RP), motor Imagery (MI), and event-related potentials (ERP). However, brain fingerprint identification that relies on external stimuli assumes that subjects can consistently receive these stimuli during the experiment without any physiological impairments, which may not always be realistic. Furthermore, much of the existing research in this field focuses on manually designed features tailored to specific tasks, which poses significant challenges in terms of generalization to real-world scenarios.

In addition to the challenges mentioned above, traditional machine learning methods often require time-consuming preprocessing of raw EEG data, which typically suffers from a low signal-to-noise ratio. During this preprocessing, valuable information is at risk of being lost. Deep learning, however, addresses this limitation by capturing both high-level features and latent dependencies directly from the raw data. Currently, deep learning has been widely applied to EEG-based Brain-Computer Interface (BCI) classification tasks. Cecotti et al. [1] were among the first to apply Convolutional Neural Networks (CNNs) to this field, although their approach yielded results that were slightly inferior to those of traditional machine learning methods. In response, some researchers have proposed improvements based on manually extracted features [2–5]. Other studies have demonstrated that deep learning techniques for EEG analysis allow the model to learn distinct high-level representations directly from raw data, offering improved generalization capabilities [6–8]. However, most complex deep learning models, with their large number of parameters, require substantial sample sizes for each class in order to train effectively. This is particularly challenging in brain fingerprint identification, where there are typically large numbers of individuals but very limited samples for each individual in real-world scenarios. Therefore, it is crucial to identify an effective method that can function well with limited sample sizes, particularly in practical applications.

In this chapter, a Convolutional Tensor-Train Neural Network (CTNN) is proposed to advance brain fingerprint identification. The CTNN effectively captures high-level local features in both the temporal and spatial domains through a deep convolutional neural network architecture, while maintaining a compact number of parameters. Given the large number of individuals involved, traditional approaches face an explosion of parameters, making them less efficient. To address this, the TensorNet, based on low-rank representation, is employed to explore high-dimensional interactions within the local features, significantly reducing the number of parameters required. This approach enables efficient multi-task brain fingerprint identification, overcoming the limitations imposed by sample size constraints.

8.2 Convolutional Tensor-Train Neural Network

8.2.1 Tensor Network

In this chapter, $\mathcal{X} \in \mathbb{R}^{I_1 \times I_2 \times \cdots \times I_d}$ represents a d-dimensional tensor with $\prod_{i=1}^{d} I_i$ elements. The element $\mathcal{X}(i_1, i_2, \ldots, i_d)$ is an entry in the tensor $\mathcal{X}$, $i_k = 1, 2, \ldots, I_k$, and $k = 1, 2, \ldots, d$.

The *TT-Format* is a specific form of the tensor $\mathcal{X} \in \mathbb{R}^{I_1 \times I_2 \times \cdots \times I_d}$ using d third-order core tensors. Each element of $\mathcal{X}$ can be expressed as a product of these core tensors $\mathcal{G}$ as follows:

$$\mathcal{X}(i_1, i_2, \ldots, i_d) = \mathcal{G}_1(i_1)\mathcal{G}_2(i_2) \cdots \mathcal{G}_d(i_d). \tag{8.1}$$

Here, $\mathcal{G}_k \in \mathbb{R}^{r_{k-1} \times I_k \times r_k}$, with $i_k \in [1, I_k]$, $\forall k \in [1, d]$, and $r_0 = r_d = 1$. The sequence $R = r_0, r_1, \ldots r_d$ represents the ranks of the tensor train, which determine the complexity of the tensor train format. Note that the first and last ranks are fixed at 1, so $\mathcal{G}_1(i_1)$ and $\mathcal{G}_d(i_d)$ are vectors, while for $k = 2, 3, \ldots, d - 1$, $\mathcal{G}_k(i_k)$ is a slice matrix with dimensions $r_{k-1} \times r_k$ from the core tensor $\mathcal{G}_k$. Ultimately, the tensor train sequence is computed as a series of vector-matrix-vector multiplications, producing a scalar output. Figure 8.1 illustrates this calculation process. In this figure, the tensor object is represented by a node, and the edges attached to the node correspond to the tensor's dimensions [9]. Additionally, the weight matrix expressed in the *TT-Format* in a fully connected layer is referred to as the *TT-Layer* [10].

8.2.2 Convolutional Tensor-Train Neural Network Architecture

As shown in Fig. 8.2, the model consists primarily of a convolutional neural network (CNN) and a TensorNet (TN).

First, the input samples, with a shape of $[C, S]$ are passed into a $2D$ convolutional layer containing P temporal kernels of size $[1, S/8]$, which performs convolution along the temporal domain for each channel, with a stride set to 1. This produces P

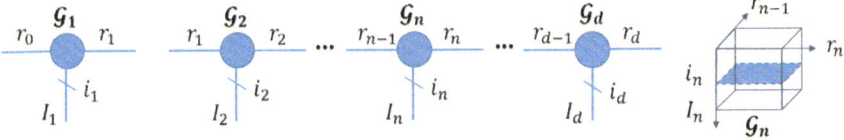

Fig. 8.1 Tensor-Train Format. The element in tensor $\mathcal{X} \in R^{I_1 \times I_2 \times \cdots \times I_d}$ can be obtained by performing a series of vector-matrix-vector multiplications on the core tensors $\mathcal{G}$. Note that $r_0 = r_d = 1$

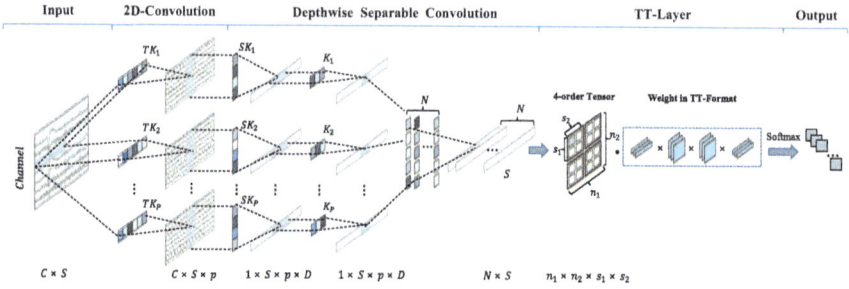

Fig. 8.2 Implementation of CTNN model for brain fingerprint identification. TK, SK, K represent temporal kernels, spatial kernels, and 1D-kernels, respectively. The bottom row illustrates the shape of the corresponding input/output. Additionally, D is used to control the number of SK to be learned for each feature map. Note that $D = 1$ is used for illustration purposes in the figure, and $S = s_1 \times s_2$, $N = n_1 \times n_2$

feature maps of size $[C, S]$. Next, the feature maps are fed into a depthwise separable convolution structure, composed of a depthwise convolution layer and a separable convolution layer. The depthwise convolution layer contains $P \times D$ spatial kernels of size $[C, 1]$, where D is a parameter that controls the number of spatial kernels for each feature map, and $D = 1$ is used in Fig. 8.2. The resulting $P \times D$ spatial feature maps, with a shape of $[1, S]$, are then passed into the separable convolution layer, where convolution operations are performed using 1D-kernels of size $[1, 16]$ and $P \times N$ point-wise kernels. Batch normalization is applied before the nonlinear activation using the Rectified Linear Unit (ReLU). A dropout probability of 0.5 is set to prevent overfitting. Finally, the CNN structure outputs N pieces of local information, each with a shape of $[1, S]$, in both the temporal and spatial domains.

Then, local features of size $[N, S]$ are input to the TT-layer, a fully connected layer with the weight matrix represented in TT-$Format$. This layer can be viewed as applying a linear transformation to an n-dimensional input vector, and its mathematical representation is given by Eq. 8.2:

$$\mathbf{y} = \mathbf{W}\mathbf{x} + \mathbf{b}. \tag{8.2}$$

As shown in Fig. 8.2 for illustration purposes, $\mathbf{W} \in \mathbb{R}^{M \times NS}$ is a large weight matrix, where M represents the number of subjects, and $\mathbf{b}$ is the bias vector. The elements of $\mathbf{y}$ can be expressed as:

$$\mathbf{y}(i) = \sum_{j=1}^{NS} \mathbf{W}(i, j)\mathbf{x}(j) + \mathbf{b}(i). \tag{8.3}$$

The principal idea of the TT-$Layer$ is to convert $\mathbf{y}$, $\mathbf{x}$ and $\mathbf{b}$ into tensor forms $\mathcal{Y}$, $\mathcal{X}$, $\mathcal{B}$, respectively. Then, the TT-$Layer$ represents the weight matrix $\mathbf{W}$ as $\mathcal{W}$ in TT-$Format$.

8.2 Convolutional Tensor-Train Neural Network

First, the local feature with size $[N, S]$ is reshaped into a 4th-order tensor $\mathcal{X} \in \mathbb{R}^{n_1 \times n_2 \times s_1 \times s_2}$, where $N = n_1 \times n_2$, and $S = s_1 \times s_2$. For example, if the shape of the local features is $[200, 64]$, it is reshaped into a 4th-order tensor with a size of $[20, 10, 8, 8]$, with each dimension balanced as much as possible. Then, it is necessary to establish the vector **y**, the bias vector **b** and the matrix **W** using 4th-order tensors.

For the vectors $\mathbf{y}, \mathbf{b} \in \mathbb{R}^M$, where $M = \prod_{k=1}^{d} m_k$, and $d = 4$, a bijective function $F(i) = (f_1(i), f_2(i), f_3(i), f_4(i)) = (i_1, i_2, i_3, i_4)$ can be established to relate the index i of the vector to the index (i_1, i_2, i_3, i_4) of 4th-order tensor $\mathcal{Y}$, and $\mathcal{B}$, where $i \in 1, 2, \ldots, M$. The elements $\mathbf{y}(i)$, and $\mathbf{b}(i)$ correspond to the entries of the vectors **y**, and **b**, respectively. The size of the k-th dimension for the $\mathcal{Y}$, and $\mathcal{B}$ is m_k. Thus, the relationship between the tensors $\mathcal{Y}$, $\mathcal{B}$ and the vector **y**, **b** is given by:

$$\mathcal{Y}(F(i)) = \mathcal{Y}(i_1, i_2, i_3, i_4) = \mathbf{y}(i),$$
$$\mathcal{B}(F(i)) = \mathcal{B}(i_1, i_2, i_3, i_4) = \mathbf{b}(i). \tag{8.4}$$

Similarly, for the weight matrix $\mathbf{W} \in \mathbb{R}^{M \times N'}$, where $M = \prod_{k=1}^{d} m_k$ and $N' = \prod_{k=1}^{d} n_k$, with $d = 4, N' = N \times S, n_3 = s_1, n_4 = s_2$, two bijective functions $F(i) = (f_1(i), f_2(i), f_3(i), f_4(i)) = (i_1, i_2, i_3, i_4)$ and $G(j) = (g_1(j), g_2(j), g_3(j), g_4(j)) = (j_1, j_2, j_3, j_4)$ are established to relate the index pair (i, j) of the matrix **W** to the index pair (i_1, i_2, i_3, i_4) and (j_1, j_2, j_3, j_4) of the tensor $\mathcal{W}$, where $\mathbf{W}(i, j)$ is an element of the matrix **W**. The size of the k-th dimension in $\mathcal{W}$ is $m_k n_k$. Combining Eq. 8.1, the relationship between the matrix **W** and the tensor $\mathcal{W}$ in $TT\text{-}Format$ is given as:

$$\begin{aligned}\mathbf{W}(i, j) &= \mathcal{W}((f_1(i), g_1(j)), \ldots, (f_4(i), g_4(j))) \\ &= \mathcal{G}_1[f_1(i), g_1(j)], \ldots, \mathcal{G}_4[f_4(i), g_4(j)] \\ &= \mathcal{G}_1[i_1, j_1], \ldots, \mathcal{G}_4[i_4, j_4]. \end{aligned} \tag{8.5}$$

Subsequently, Eq. 8.3, based on the matrix **W** expressed in $TT\text{-}Format$, can be rewrited as Eq. 8.6:

$$\mathcal{Y}(i_1, \ldots, i_4) = \sum_{j_1, \ldots, j_4=1}^{n_1, \ldots, n_4} \mathcal{G}_1[i_1, j_1], \ldots, \mathcal{G}_4[i_4, j_4] \cdot$$
$$\mathcal{X}(j_1, \ldots, j_4) + \mathcal{B}(i_1, \ldots, i_4). \tag{8.6}$$

Local features are transformed into high-order tensors in the $TT\text{-}Layer$, where high-dimensional potential dependencies are explored through the weight matrix represented in low-rank $TT\text{-}Format$. This process integrates the local information into a global representation. Finally, the global features are passed directly to the Softmax activation function for classification.

8.3 Brain Fingerprint Identification with CTNN

8.3.1 EEG Datasets

For brain fingerprint identification, several single-task EEG datasets are utilized, including the EEG Motor Movement/Imagery Dataset (EEGMMI), the DEAP dataset, and a self-collected multi-task EEG dataset (MTED). The EEGMMI dataset consists of EEG recordings collected during motor movement and motor imagery tasks, making it highly suitable for exploring brain fingerprint identification related to motor cognitive states. The DEAP dataset, on the other hand, contains EEG data obtained while participants watched emotional videos, providing insight into the brain's response to emotional stimuli and offering a different context for brain fingerprint identification. In addition to these well-established datasets, we also incorporate the MTED, a self-collected multi-task EEG dataset. This dataset is specifically designed to simulate more complex, real-world scenarios where individuals perform multiple cognitive tasks, allowing for the evaluation of brain fingerprint identification in conditions that involve diverse mental states. These datasets provide a wide variety of EEG data, which is essential for testing the robustness and accuracy of the proposed model in real-world conditions. A detailed description of each dataset used in the experiments is provided in Table 8.1.

8.3.2 Data Pre-processing

This section details the preprocessing steps applied to the three datasets used for brain fingerprint identification. We then provide a comprehensive description of the CTNN architecture proposed in this chapter. As shown in Fig. 8.3, the most relevant channels are identified and extracted. For signal preprocessing, a 2nd-order Butterworth bandpass filter is applied to isolate the 4–45 Hz frequency range from the EEG signals in each dataset. The filtered data is then segmented into 1-second trials, ensuring consistent temporal resolution across all samples. Finally, these trials are normalized and reshaped into matrices of size $C \times S$, where C represents the number of channels and S corresponds to the number of data points. These matrices

Table 8.1 Details of the four datasets

Dataset	Sampling rate	Channel	Subject	Sample
EEGMMI-RS	160	32	105	120
EEGMMI-MI	160	32	105	490
MTED	200	32	20	520
DEAP	128	32	32	2400

8.3 Brain Fingerprint Identification with CTNN

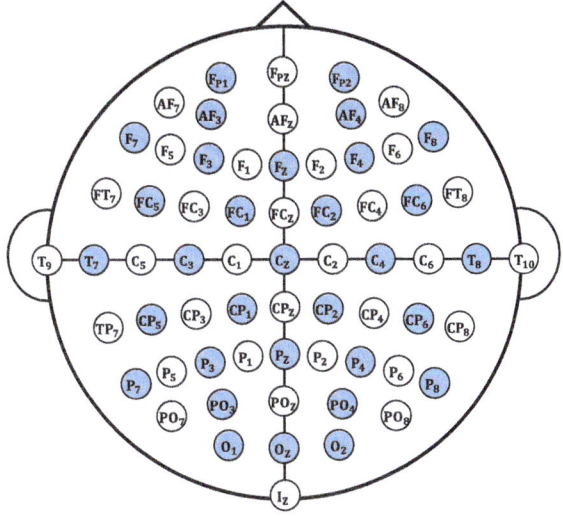

Fig. 8.3 The positions of channels. The blue-highlighted channels indicate those used in the experiments

form the input to the convolutional layer, allowing the model to efficiently learn discriminative features for brain fingerprint identification.

8.3.3 Experiment Setup

In this section, CNN [11] and CNN-RNN [12] are introduced as baseline models, both of which are widely used in EEG-based biometric identification tasks. The CNN architecture consists of two convolutional layers and two fully connected layers. In the CNN-RNN model, a convolutional neural network is followed by a recurrent layer, using either LSTM or GRU units. The structure of the model for MTED is outlined in Table 8.2. To evaluate the model performance, several metrics are employed: accuracy (ACC), F1 score (F1), mean absolute error (MAE), and equal error rate (EER). Experimental evaluations are conducted using fivefold cross-validation, with test samples comprising 20% of the total dataset.

8.3.4 Comparison of the Performance on Different Models

In this section, the TT-Ranks of CTNN are set to [1, 5, 5, 5, 1] for the experiments. The results are compared with CNN, CNN-LSTM, CNN-GRU, and TN with varying ranks, as shown in Table 8.3. The bottom row of the table records the performance of CTNN. CTNN demonstrates superior performance across most metrics, achieving an ACC of over 99% and an EER below 0.1% on both single-task and multi-task

Table 8.2 Detailed comparison for the structure of models used in the experiment

Model	Layer	Output	Kernel	Parameter	Total
CNN [11]	Input	32×200	–	–	4620872
	Convlution2D	$32 \times 200 \times 16$	$32 \times 1(16)$	512	
	Convlution2D	$32 \times 181 \times 32$	$1 \times 20(32)$	10240	
	max_pooling	$16 \times 90 \times 32$	–	–	
	Dense	100	–	4608100	
	Dense (Softmax)	20	–	2020	
CNN-RNN [12]	Input	32×200	–	–	$1104788(GRU)$
					$1125332(LSTM)$
	Convlution2D[a]	$5 \times 9 \times 5 \times 128$	$3 \times 3(128)$	230528	
	Convlution2D[a]	$5 \times 9 \times 5 \times 64$	$3 \times 3(64)$	73792	
	Dense[a]	5×256	–	737536	
	RNN (LSTM/GRU)	64	–	$61632(GRU)$ $82176(LSTM)$	
	Dense (Softmax)	20	–	1300	
TN [10], rank = 9	TT-Layer	256	–	4644	4932
	TT-Layer (Softmax)	20	–	288	
CTNN	Input	32×200	–	–	4600
	Convlution2D	$32 \times 200 \times 8$	$1 \times 20(8)$	200	
	Depthwise convolution	$1 \times 200 \times 16$	$32 \times 1(8 \times 2)$	512	
	Separable convolution	$1 \times 200 \times 64$	$1 \times 16(8 \times 2)$ $1 \times 1(8 \times 2 \times 64)$	1280	
	TT-Layer	256	–	2260	
	TT-Layer (Softmax)	20	–	348	

[a] Indicates the structure with time distributed

datasets, effectively addressing the limitations of single-task models. Furthermore, the results in Table 8.3 show that the performance of CNN and CNN-GRU approaches that of CTNN in MTED and DEAP datasets, respectively.

Compared to CNN, CTNN achieves better performance because it applies high-dimensional tensorization to the local features obtained through convolution. This allows the model to capture more extensive high-order information and better explore multilinear intercorrelations. As a result, the extracted brainprint features are more authentic and efficient for recognition. Furthermore, compared to traditional TNs, CTNN benefits from integrating CNN layers directly with raw EEG data. This integration enables the extraction of high-level dependencies, including those related to various frequency bands and spatial characteristics. Consequently, the data used in tensorization is more relevant to brainprint features. By leveraging the strengths of both CNN and TN architectures, CTNN effectively performs multi-task brain fingerprint identification.

8.3 Brain Fingerprint Identification with CTNN

Table 8.3 Comparison of the performance on different models

Models	EEGMMI-RS				EEGMMI-MI				MTED				DEAP			
	ACC	F1	MAE	EER	ACC	F1	MAE	EER	ACC	F1	MAE	EER	ACC	F1	MAE	EER
CNN [11]	**87.83**	**86.26**	**4.53**	**2.75**	**96.74**	**96.69**	**1.04**	1.64	**98.92**	**98.89**	**0.08**	1.29	96.06	96.58	0.31	3.06
CNN-LSTM [12]	57.74	51.26	16.83	7.29	89.05	87.80	4.41	1.77	95.81	94.47	0.50	2.11	99.38	99.39	0.06	0.24
CNN-GRU [12]	57.47	52.65	15.52	6.23	86.50	84.86	5.68	**1.57**	97.41	97.01	0.25	1.45	**99.83**	**99.84**	**0.02**	**0.08**
TN [10], rank = 1	8.84	9.44	29.76	29.69	5.38	4.91	31.39	34.07	73.19	72.51	2.42	5.00	79.27	76.57	1.87	6.46
TN, rank = 3	20.57	20.02	25.88	25.81	17.75	17.70	24.86	28.31	94.89	94.67	0.42	1.33	92.34	91.45	0.93	2.61
TN, rank = 5	27.50	27.90	24.41	24.10	31.87	31.85	20.90	21.86	94.00	93.66	0.52	1.39	97.59	97.44	0.32	1.12
TN, rank = 7	28.59	28.50	23.75	20.79	36.28	36.66	18.95	17.22	97.73	97.69	0.19	**0.52**	98.90	98.83	0.13	0.36
TN, rank = 9	35.42	34.80	21.80	20.52	44.60	44.60	17.27	14.76	95.72	95.58	0.34	1.08	98.84	98.83	0.16	0.29
CTNN	99.53	99.50	0.19	0.039	99.74	99.74	0.089	0.048	99.88	99.87	0.009	0.018	100.00	100.00	0.000	0.000

8.3.5 Comparison of Multi-task Brain Fingerprint Identification with Different TT-Ranks

As TT-Rank plays a crucial role in TT-Layer, we investigated how CTNN performs under different TT-Ranks. In this section, the rank R varies from $[1, 1, 1, 1, 1]$ to $[1, 9, 9, 9, 1]$. As shown in Fig. 8.4, CTNN achieves an accuracy of over 95% on EEGMMI-RS even at rank-1 across the four datasets. With ranks of 2 or 3, CTNN's accuracy consistently exceeds 99% on EEGMMI-RS, EEGMMI-MI, and MTED. For DEAP, the model reaches 99% accuracy even at rank-1. By analyzing the results presented in Tables 8.1 and 8.3, we observe an intriguing pattern: as the number of samples per class decreases, the performance of baseline models gradually declines. In contrast, CTNN demonstrates robust generalization. When dealing with scenarios involving a large number of subjects but only a small number of samples per subject, such as EEGMMI-RS, CTNN shows clear advantages.

The TT-Rank directly controls the number of parameters in the TT-Layer. In real-world applications with numerous subjects, traditional methods often result in a significant computational burden. For example, if the features extracted by CNN have a shape of [200, 64], a fully connected layer used to classify 100 participants would

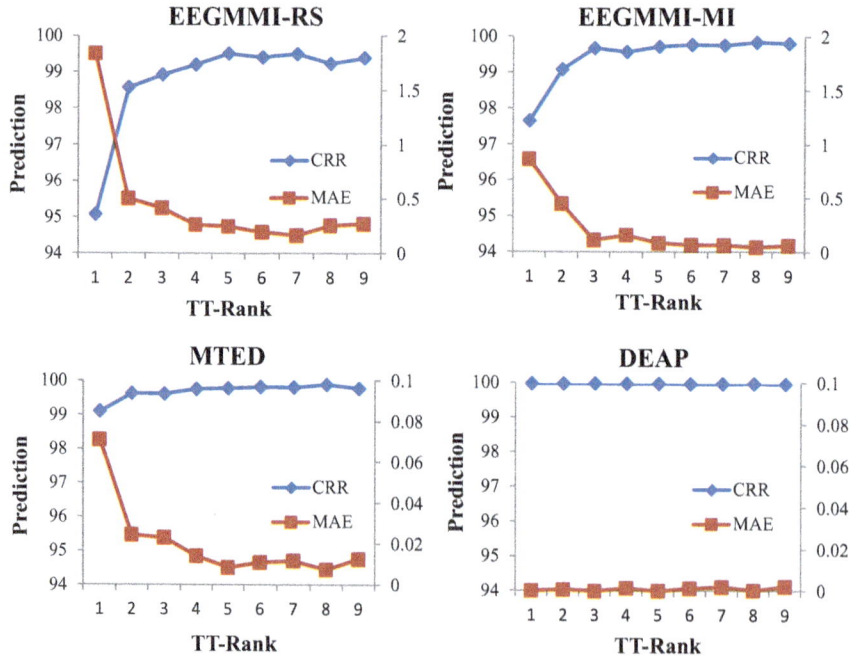

Fig. 8.4 The effect of different TT-Ranks on the performance of CTNN for brain fingerprint identification across four datasets: EEGMMI-RS, EEGMMI-MI, MTED, and DEAP. The blue line represents ACC, and the red line represents MAE

require training $200 \times 64 \times 100 = 1280000$ parameters. In contrast, a TT-Layer with the TT-Rank of $[1, 5, 5, 5, 1]$ would require only $20 \times 5 \times 1 \times 5 + 10 \times 2 \times 5 \times 5 + 8 \times 2 \times 5 \times 5 + 8 \times 5 \times 5 \times 1 = 1600$ parameters, which is 800 times fewer than the fully connected layer. This low-rank representation effectively captures high-dimensional interactions while maintaining a lightweight parameter count. As a result, CTNN scales well to datasets with limited training samples, making it a practical solution for real-world scenarios.

8.3.6 Comparison of Multi-task Brain Fingerprint Identification Among EEG from Different Frequency Bands

EEG signals arise from spontaneous, rhythmic neural activity that is typically categorized into Theta (4–8 Hz), Alpha (8–15 Hz), Beta (15–32 Hz), and Gamma (32–45 Hz) frequency bands. In this section, we investigate the impact of different frequency bands on brain fingerprint identification. Using a Butterworth bandpass filter, these frequency components are extracted from four datasets, and the TT-rank is fixed at $[1, 5, 5, 5, 1]$ for the experiments. The analysis, based on the results shown in Table 8.4, demonstrates that the accuracy (ACC) in the 4–8 Hz band is the lowest, while the highest accuracy is achieved in the higher frequency bands (15–45 Hz). Furthermore, the results from the full frequency range (4–45 Hz) indicate that it can deliver comparable or even the best performance.

This suggests that brainprint information may be distributed across multiple frequency bands. It is plausible to infer that the accuracy of brain fingerprint identification tends to be lower in low-frequency bands compared to high-frequency bands, given that the subjects in the dataset demonstrate robust cognitive abilities.

8.3.7 Comparison of Multi-task Brain Fingerprint Identification with Different Channels Selections

Currently, most studies on brain fingerprint identification using a small number of channels do not specify the method for channel selection. In this section, the four groups ($D = 4$) of spatial kernels are visualized separately across the four datasets, as shown in Fig. 8.5. Analyzing the visualized results in Fig. 8.6, it is evident that brainprint features are primarily concentrated around Fz, FC1, FC2, Cz, CP1, CP2, and Pz. Building on these observations, as shown in Fig. 8.7, four sets of channels are selected for performance comparison and statistical analysis. The corresponding results are presented in Table 8.5, Figs. 8.8, and 8.9. As illustrated in Figs. 8.8 and 8.9, the performance of the selected 7 channels and 32 channels across the four datasets is comparable, with no significant differences observed ($p > 0.05$).

Table 8.4 Comparison with different frequency bands for brain fingerprint identification

Frequncy	EEGMMI-RS				EEGMMI-MI				MTED				DEAP			
	ACC	F1	MAE	EER	ACC	F1	MAE	EER	ACC	F1	MAE	EER	ACC	F1	MAE	EER
4–8hz	97.50	97.32	0.856	0.357	98.49	98.30	0.620	0.232	98.58	98.52	0.092	0.280	99.93	99.92	0.014	0.001
8–15hz	98.99	98.94	0.366	0.227	99.38	99.37	0.233	0.087	99.43	99.40	0.041	0.029	99.99	99.99	0.001	0.001
15–32hz	99.38	99.33	0.228	*0.040*	*99.60*	*99.59*	*0.139*	**0.048**	**99.92**	**99.92**	**0.005**	0.020	**100.00**	**100.00**	**0.000**	**0.000**
32–45hz	**99.61**	**99.59**	**0.155**	*0.040*	99.48	99.44	0.159	0.058	99.83	99.82	0.012	**0.010**	99.99	99.99	0.001	0.006
4–45 Hz	*99.53*	*99.50*	*0.190*	**0.039**	**99.74**	**99.74**	**0.089**	**0.048**	*99.88*	*99.87*	*0.009*	*0.018*	**100.00**	**100.00**	**0.000**	**0.000**

The highlight in black represents the best-performing frequency band, while the second-best performance is indicated by the italicized underscores

8.3 Brain Fingerprint Identification with CTNN

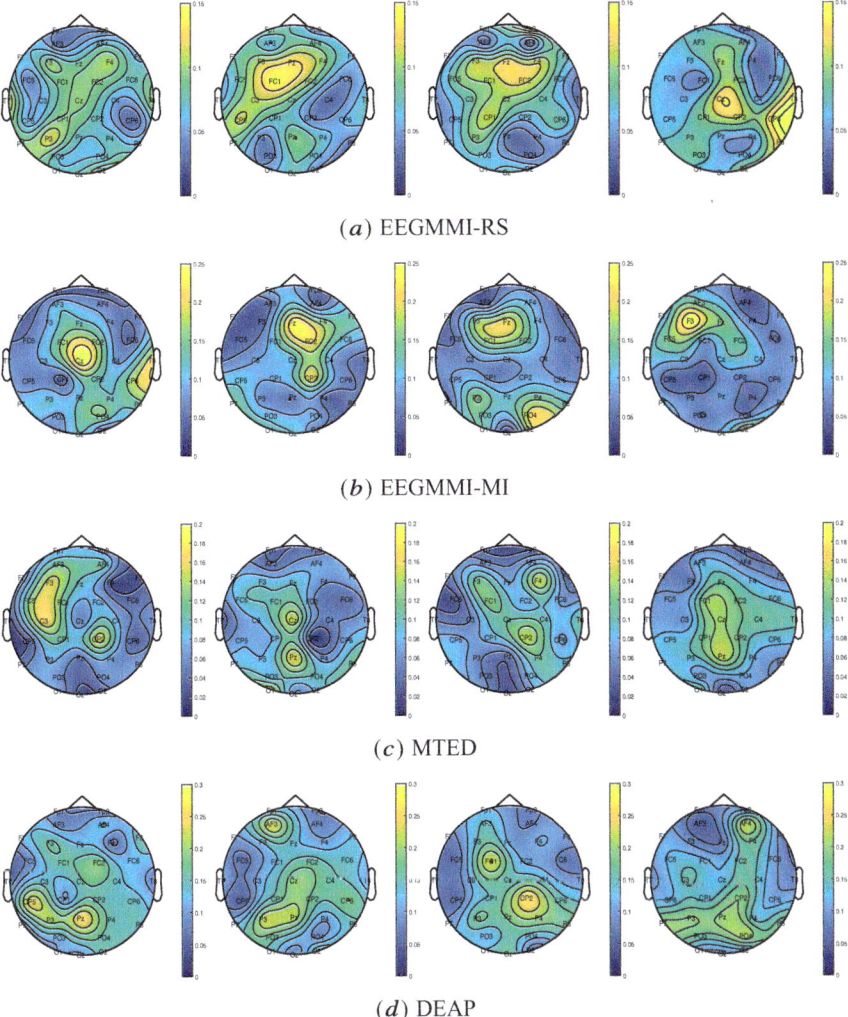

Fig. 8.5 Visualization of spatial kernels on EEGMMI-RS, EEGMMI-MI, MTED, and DEAP. Brainprint features are predominantly concentrated around the Fz, Cz, and Pz regions. Channels with higher corresponding weights are represented with colors closer to yellow

These findings indicate that brain fingerprint identification relies on a select set of EEG channels. The CTNN model successfully extracts information crucial for brain fingerprint identification, achieving strong performance even when limited to a small number of channels.

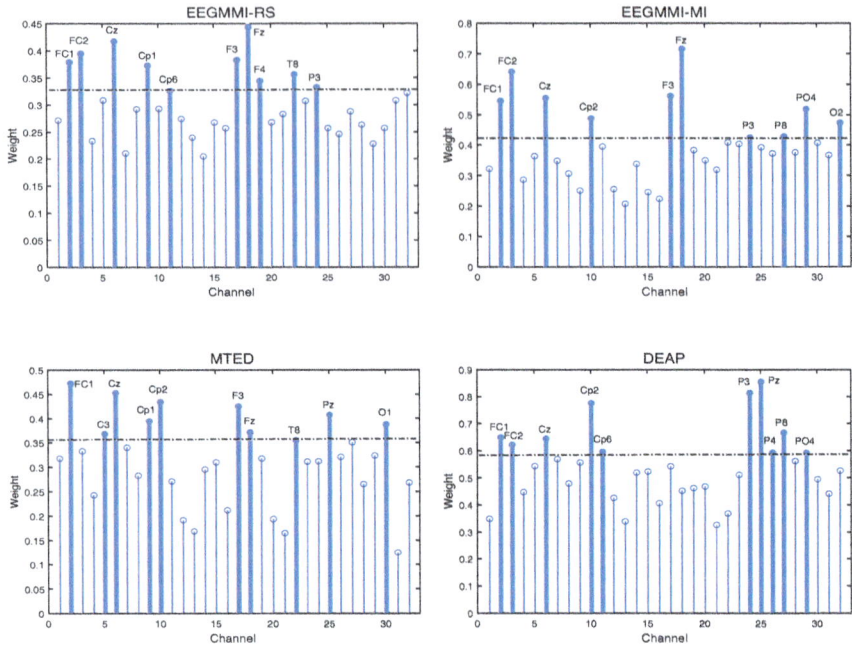

Fig. 8.6 The channel weights for the four datasets are tabulated. The channels above the dotted line are the ten with the highest weights. Among these, FC1, FC2, Cz, CP1, CP2, Fz, and Pz consistently show relatively high weights across multiple datasets

8.4 Conclusion

This chapter introduced the CTNN model, a hybrid approach that combines convolutional neural networks and TensorNet to address the challenges of multi-task brain fingerprint identification. By employing depthwise separable convolution to capture latent dependencies and a low-rank tensor train representation to reduce parameter complexity, CTNN offers a highly efficient framework for extracting and leveraging the high-order interactions embedded within EEG data.

One of CTNN's notable strengths is its ability to maintain high recognition accuracy despite a limited number of training samples. This capability is especially critical in real-world scenarios where large subject pools often accompany small per subject sample sizes. Unlike traditional approaches that require extensive datasets or are confined to single-task settings, CTNN successfully integrates multi-task information, broadening its applicability across diverse EEG-based identity recognition tasks.

Another significant advantage of CTNN lies in its efficient parameterization. The low-rank tensor train approach minimizes the need for a large number of parameters, which in turn reduces computational requirements. This efficiency makes CTNN particularly attractive for real-world applications, such as portable EEG systems or environments with limited hardware resources. Moreover, the model's reliance

8.4 Conclusion

Table 8.5 Performance comparison and statistical analysis of different channel sets.

Channels	EEGMMI-RS				EEGMMI-MI				MTED				DEAP			
	ACC	SD	EER	P-value	ACC	SD	EER	P-value	ACC	SD	EER	P-value	ACC	SD	EER	P-value
3	87.59	0.32	1.11	0.0000	90.23	0.69	0.96	0.0000	96.83	0.69	0.83	0.0000	97.77	0.99	0.60	0.0010
5	97.12	0.4	0.29	0.0032	98.01	0.29	0.33	0.0051	98.98	0.37	0.10	0.0053	99.79	0.05	0.03	0.0000
6	98.02	0.3	0.30	0.0371	98.3	0.24	0.27	0.0150	99.3	0.11	0.05	0.0187	99.98	0.02	0.01	0.0665△
7	**98.49**	**0.15**	**0.20**	**0.1502△**	**98.72**	**0.33**	**0.21**	**0.1022△**	**99.47**	**0.18**	**0.08**	**0.0814△**	**99.99**	**0.01**	**0.01**	**0.3302△**

1 Triangles are defined as follows $p > 0.05$
2 SD: Standard Deviation

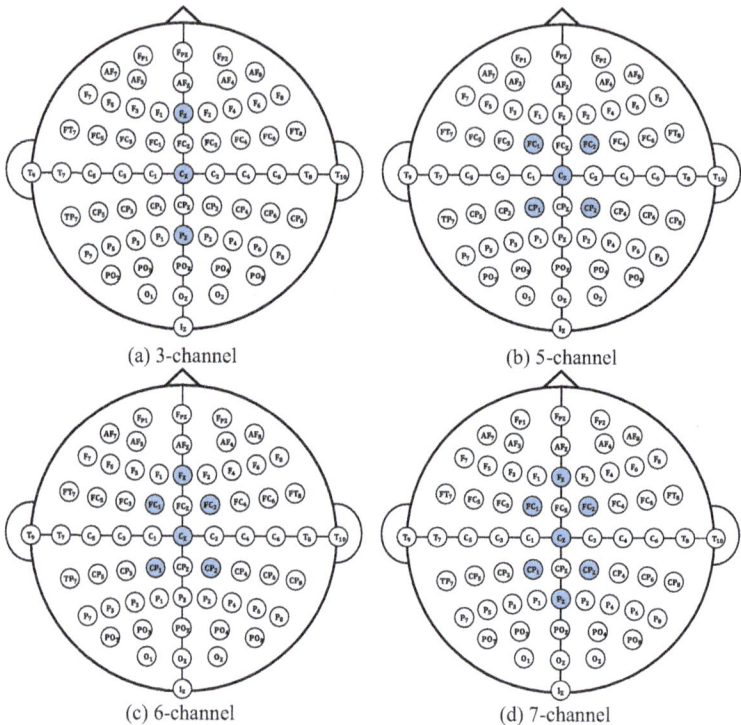

Fig. 8.7 Various channel selections for multi-task brain fingerprint identification. The blue-highlighted channels indicate those chosen for this experiment

on high-frequency EEG bands (8–45 Hz) and its ability to pinpoint specific spatial regions, centered on Cz and a select group of dominant channels, highlight its capacity to isolate critical brainprint features. These characteristics enable more focused data collection and preprocessing strategies, further enhancing the overall effectiveness of the system.

However, the CTNN model is not without limitations. Currently, the TT ranks must be manually specified, which can limit adaptability and add complexity for non-expert users. Overcoming this challenge will likely involve developing methods for automated or dynamically adjustable rank selection, enabling the model to better adapt to varying datasets and conditions. Additionally, exploring more generalized approaches to handle diverse subject populations and varying experimental conditions will be critical for broadening the applicability of CTNN.

In conclusion, CTNN represents a promising step forward in brain fingerprint identification, balancing accuracy, efficiency, and flexibility. By addressing its current limitations and continuing to refine its design, CTNN has the potential to become a highly versatile tool for EEG-based identity recognition, paving the way for more accessible, portable, and real-world applications.

8.4 Conclusion

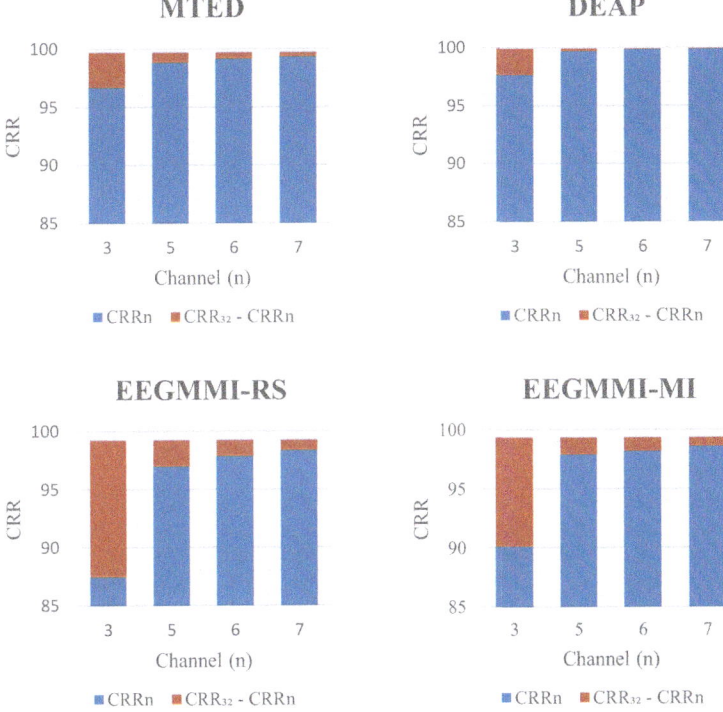

Fig. 8.8 Performance on various channel subsets is presented. Blue represents the performance of the different channel sets, while red shows the performance differences between these subsets and the full 32-channel configuration. For the experiments, the TT-rank of CTNN is set as [1, 7, 7, 7, 1], [1, 7, 7, 7, 1], [1, 8, 8, 8, 1], and [1, 5, 5, 5, 1] for EEGMMI-RS, EEGMMI-MI, MTED, and DEAP respectively

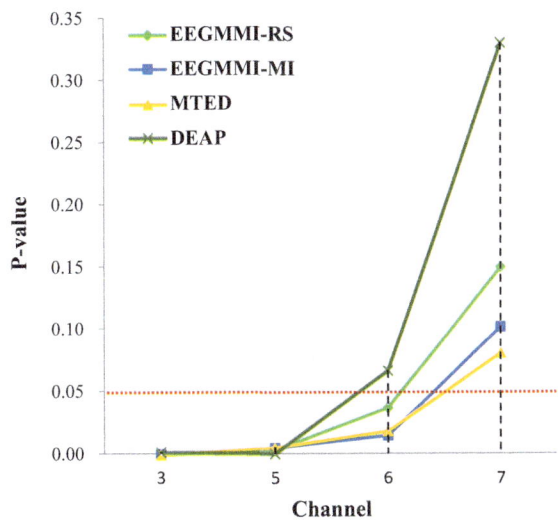

Fig. 8.9 p-value analysis for ACC of various channel sets compared to 32 channels. The red line indicates the 0.05 threshold, below which the difference is considered statistically significant ($p < 0.05$)

References

1. Cecotti H, Graser A (2010) Convolutional neural networks for P300 detection with application to brain-computer interfaces. IEEE Trans Pattern Anal Mach Intell 33(3):433–445
2. Na L, Li T, Ren X, Miao H (2016) A deep learning scheme for motor imagery classification based on restricted Boltzmann machines. IEEE Trans Neural Syst Rehabil Eng 25(6):566–576
3. Sturm I, Lapuschkin S, Samek W, Müller K-R (2016) Interpretable deep neural networks for single-trial EEG classification. J Neurosci Methods 274:141–145
4. Tabar YR, Halici U (2016) A novel deep learning approach for classification of EEG motor imagery signals. J Neural Eng 14(1):016003
5. Kwak NS, Müller KR, Lee SW (2017) A convolutional neural network for steady state visual evoked potential classification under ambulatory environment. PloS One 12(2):e0172578
6. Schirrmeister RT, Springenberg JT, Fiederer LD, Glasstetter M, Eggensperger K, Tangermann M, Hutter F, Burgard W, Ball T (2017) Deep learning with convolutional neural networks for EEG decoding and visualization. Hum Brain Mapp 38(11):5391–5420
7. Lawhern VJ, Solon AJ, Waytowich NR, Gordon SM, Hung CP, Lance BJ (2018) EEGNet: a compact convolutional neural network for EEG-based brain–computer interfaces. J Neural Eng 15(5):056013
8. Zhang X, Yao L, Wang X, Monaghan J, Mcalpine D, Zhang Y (2019) A survey on deep learning based brain computer interface: recent advances and new frontiers. arXiv:1905.04149
9. Cichocki A, Lee N, Oseledets I, Phan AH, Zhao Q, Mandic DP (2016) Tensor networks for dimensionality reduction and large-scale optimization: Part 1 low-rank tensor decompositions. Found Trends® Mach Learn 9(4–5):249–429
10. Novikov A, Podoprikhin D, Osokin A, Vetrov DP (2015) Tensorizing neural networks. In: Advances in neural information processing systems, pp 442–450
11. Mao Z, Yao WX, Huang Y (2017) EEG-based biometric identification with deep learning. In: 2017 8th International IEEE/EMBS conference on neural engineering (NER). IEEE, pp 609–612
12. Wilaiprasitporn T, Ditthapron A, Matchaparn K, Tongbuasirilai T, Banluesombatkul N, Chuangsuwanich E (2020) Affective EEG-based person identification using the deep learning approach. IEEE Trans Cogn Dev Syst 12(3):486–496

Open Access This chapter is licensed under the terms of the Creative Commons Attribution-NonCommercial-NoDerivatives 4.0 International License (http://creativecommons.org/licenses/by-nc-nd/4.0/), which permits any noncommercial use, sharing, distribution and reproduction in any medium or format, as long as you give appropriate credit to the original author(s) and the source, provide a link to the Creative Commons license and indicate if you modified the licensed material. You do not have permission under this license to share adapted material derived from this chapter or parts of it.

The images or other third party material in this chapter are included in the chapter's Creative Commons license, unless indicated otherwise in a credit line to the material. If material is not included in the chapter's Creative Commons license and your intended use is not permitted by statutory regulation or exceeds the permitted use, you will need to obtain permission directly from the copyright holder.

Chapter 9
Multi-task and Multi-session Brain Fingerprint Identification with Attention Neural Network with Domain Adaptation Learning

Abstract EEG signals hold great promise as biometric identifiers due to their invisibility and adaptability to high-security application scenarios. However, extracting reliable EEG identity features remains a challenge, primarily due to interference from device-related variations and the inherent differences in the subject's state across multiple sessions. Current methods often treat each training session as a separate domain, which is problematic due to the differing data distributions across sessions. While many multi-source domain adaptation techniques attempt to bridge the domain gap between multiple source and target domains individually, they fail to account for the interrelationships between domain-invariant features during distribution alignment. In this chapter, we propose a novel multi-source domain adaptation framework, the Tensorized Spatial-frequency Attention Network (TSFAN), designed to enhance the performance of EEG-based brain fingerprint identification in the target domain. Specifically, TSFAN models the significant relationships between domain-invariant features using a tensorized attention mechanism. This mechanism effectively incorporates appropriate spatial-frequency representations from both pairwise source-target domains as well as cross-source domains, all while mitigating the impact of distribution discrepancies among the source domains. To address the issue of dimensionality, TSFAN is further approximated in the Tucker format. By leveraging the low-rank properties of the Tucker decomposition, TSFAN is able to scale linearly with the number of domains, offering significant flexibility for extension to scenarios involving any number of sessions. Extensive experiments conducted on representative benchmark datasets demonstrate that TSFAN outperforms state-of-the-art methods, achieving superior results in terms of identification accuracy. Furthermore, our electrode selection analysis reveals that brainprint features across different sessions are distributed across various brain regions. Notably, a selection of 20 electrodes, based on the standard 10–20 system, proves to be sufficient for extracting stable and reliable identity information.

9.1 Introduction

Reliable and stable EEG identity features are fundamental to the success of EEG-based biometric systems. Traditional machine learning methods, while widely adopted in previous research, often require substantial domain expertise for effective feature extraction and are generally insufficient for achieving optimal performance [1]. In contrast, deep learning approaches have attracted significant attention in recent years due to their capability to automatically capture high-level representations and latent dependencies from complex data. Various deep learning techniques, such as Convolutional Neural Networks (CNN) [2, 3], Recurrent Neural Networks (RNN) [4], and Graph Convolutional Neural Networks (GCNN) [5], have been shown to effectively extract temporal, frequency, and spatial identity-discriminative features from EEG signals.

Despite these advances, EEG signals are inherently unstable across different sessions due to various factors, including electrode impedance, minor displacements, and fluctuations in the subject's mental or physical state [6]. This instability remains a significant obstacle to the practical application of EEG-based biometrics in real-world scenarios. Many existing studies focus either on single-session acquisitions or aggregate data from multiple training sessions, often neglecting the variations in data distributions across sessions. The challenge becomes particularly pronounced when attempting to extract domain-invariant representations, as even a single source (training session) and target (test session) domain may exhibit substantial distribution shifts. As the number of source domains increases, these shifts can result in a marked degradation in system performance.

To address this issue, several multi-source domain adaptation methods have been proposed for EEG signals [7, 8], each aiming to reduce the domain discrepancy between source and target domains individually. Although the domain-invariant features captured across different source domains can provide stable information from multiple perspectives and facilitate the transfer of more relevant information to the target domain [9], the features extracted through individual distribution alignments are often biased by the specific characteristics of the source domains. Furthermore, these methods fail to exploit the common relationships shared across multiple source domains. Therefore, it is of paramount importance to capture deep interactions between domains and uncover the shared knowledge across multiple source views in order to facilitate the extraction of robust, session-independent features.

In response to these challenges, the tensorized spatial-frequency attention network (TSFAN), a method based on domain adaptation technique, is proposed to capture the stable EEG identity features across multiple sessions. Specifically, TSFAN first maps the data from each pair of source and target domains into separate temporal feature spaces. The core innovation of TSFAN lies in its tensor-based attention mechanism, which tensorizes the spatial-frequency attention between pairwise source and target domains to extract domain-invariant spatial-frequency features. This mechanism naturally facilitates both intra-source information transfer and the modeling of complex inter-source interactions. To address the issue of dimensionality and ensure scalability, the low-rank Tucker decomposition is utilized to allow our TSFAN to scale linearly with the number of source domains. The main contributions of this chapter are summarized as follows:

- This chapter introduces the TSFAN, a method that simultaneously captures intra-source transferable information and cross-source interactions of domain-invariant features, mitigating the loss of discriminative power caused by global distribution alignment across sessions.
- A tensor-based attention is proposed, which tensorizes domain-specific attention within a low-rank Tucker Network, enabling efficient interaction between multiple source domains while avoiding the curse of dimensionality.
- Comprehensive experiments are conducted on representative benchmark datasets, demonstrating the superiority of the proposed method in terms of performance.
- Additionally, an interpretable analysis is provided for the selection of 20 electrodes based on the standard 10–20 system,, which effectively captures stable and discriminative EEG identity features across sessions.

9.2 Tensorized Spatial-Frequency Attention Network

The overall architecture of the TSFAN, as depicted in Fig. 9.1, consists of two key modules: (1) Intra-source transferable feature learning, which employs a multi-scale temporal feature extractor and a spatial-frequency feature extractor enhanced by multi-source domain adaptation to capture domain-invariant features from each pairwise source and target domain; (2) Tensorized spatial-frequency attention mechanism, which is introduced to model the complex inter-source interactions across domains. The design of this architecture is carefully crafted to effectively capture stable EEG identity features that are consistent across different sessions. The input samples are initially passed through the temporal feature extractors $\{F_t^j(\cdot)\}_{j=1}^N$ to gen-

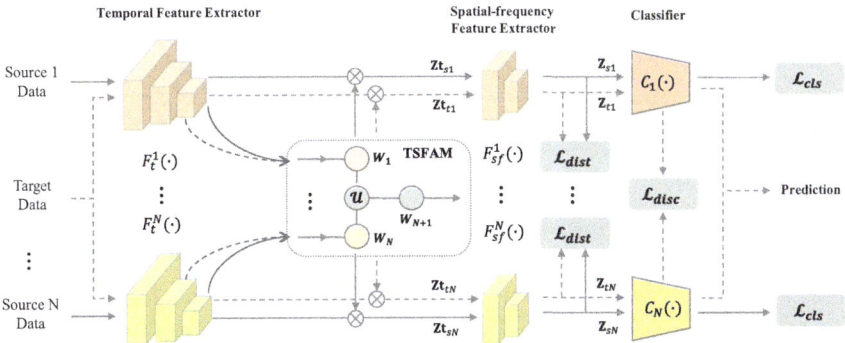

Fig. 9.1 The framework of tensorized spatial-frequency attention network (TSFAN)

Algorithm 9.1 Tensorized Spatial-Frequency Attention Network (TSFAN)

Input: N labeled source domains D_S, an unlabeled target domain D_T, batch size n, number of training iterations T;

Output: well-trained temporal feature extractors $\{F_t^j(\cdot)\}_{j=1}^{N}$, spatial-frequency feature extractors $\{F_{sf}^j(\cdot)\}_{j=1}^{N}$, tensorized spatial-frequency attention mechanism comprising $\{F_d^j(\cdot)\}_{j=1}^{N}$ and $F_b(\cdot)$, identify classifier $\{C_j(\cdot)\}_{j=1}^{N}$;

1: **for** each $t \in [1, T]$ **do**
2: Sample a mini-batch of training samples $\{(x_i^{sj}, y_i^{sj})\}_{i=1}^{n}$ from source domains D_j;
3: Sample a mini-batch of training samples $\{x_i^t\}_{i=1}^{n}$ from target domains D_t;
4: Feed $\{(x_i^{sj}, y_i^{sj})\}_{i=1}^{n}$ and $\{x_i^t\}_{i=1}^{n}$ into temporal feature extractors $F_t^j(\cdot)$ to obtain the domain-specific temporal representations $\mathbf{Zt'}_{sj}$ and $\mathbf{Zt'}_{tj}$;
5: Feed $\mathbf{Zt'}_{sj}$ and $\mathbf{Zt'}_{tj}$ into the tensorized spatial-frequency attention mechanism $F_d^j(\cdot)$ and $F_b(\cdot)$. The domain-specific temporal representations and interaction of multi-view domain-invariant features are combined as Eq. (9.8) and (9.9), resulting in the temporal representations $\mathbf{Zt}_{sj}$ and $\mathbf{Zt}_{tj}$;
6: Feed $\mathbf{Zt}_{sj}$ and $\mathbf{Zt}_{tj}$ into the spatial-frequency feature extractor $F_{sf}^j(\cdot)$ to obtain the domain-invariant identify features $\mathbf{Z}_{sj}$ and $\mathbf{Z}_{tj}$;
7: Feed the domain-invariant identify features $\mathbf{Z}_{sj}$ and $\mathbf{Z}_{tj}$ into the identify classifier $C_j(\cdot)$ for prediction;
8: Calculate the cross-entropy loss using $C_j(\mathbf{Z}_{sj})$ and $C_j(\mathbf{Z}_{tj})$ with Eq. (9.1);
9: Calculate the discrepancy loss using $C_j(\mathbf{Z}_{sj})$ and $C_j(\mathbf{Z}_{tj})$ with Eq. (9.2);
10: Calculate distribution distance loss using $\mathbf{Z}_{sj}$ and $\mathbf{Z}_{tj}$ with Eq. (9.3);
11: Update the parameters of $\{F_t^j(\cdot)\}_{j=1}^{N}$, $\{F_{sf}^j(\cdot)\}_{j=1}^{N}$, $\{F_d^j(\cdot)\}_{j=1}^{N}$, $F_b(\cdot)$, $\{C_j(\cdot)\}_{j=1}^{N}$ based on the loss function in Eq. (9.4);
12: **end for**

erate the domain-specific temporal representations. These representations are then fed into the Tensorized Spatial-Frequency Attention Mechanism (TSFAM) to obtain the temporal representations $\mathbf{Zt}_{sj}$ and $\mathbf{Zt}_{tj}$. Subsequently, these temporal representations are input into the spatial-frequency feature extractor $F_{sf}^j(\cdot)$, which produces the domain-invariant identity features $\mathbf{Z}_{sj}$ and $\mathbf{Z}_{tj}$ for the classification task. It is important to note that the TSFAM, represented in Tucker format, is briefly described using a graphical representation. In this representation, tensor objects are depicted as nodes, and each outgoing line from a node corresponds to the indices of a specific mode. The entire process is succinctly summarized in Algorithm 9.1.

9.2.1 Preliminaries

Multi-source unsupervised domain adaptation. Let $D_S = \{D_1, D_2, \ldots, D_N\}$ represent the set of N labeled source domains, each characterized by a distinct data distribution $\{p_{sj}(x, y)\}_{j=1}^{N}$. Assume that $\{X_{sj}, Y_{sj}\}_{j=1}^{N}$ denote the data from these source domains, where $X_{sj} = \{x_i^{sj}\}_{i=1}^{|X_{sj}|}$ is the collection of samples from the jth source domain, and Y_{sj} is the corresponding set of labels. Similarly, $X_t = \{x_i^t\}_{i=1}^{|X_t|}$

represents the data from the target domain D_T. The objective of our model is to predict the labels Y_t for the target domain.

Tensor Network. Multi-dimensional arrays, also known as N-way arrays, can be represented as tensors. Specifically, vectors (1st-order tensor) and matrices (2nd-order tensor) are denoted by **v** and **M**, respectively. Let $\mathcal{X} \in \mathbb{R}^{I_1 \times I_2 \times \cdots \times I_d}$ represents a d-order, where $\prod_{i=1}^{d} I_i$ is the total number of elements in the tensor. An element of the tensor $\mathcal{X}$ is donated as $x_{i_1, i_2, \ldots, i_d} = \mathcal{X}(i_1, i_2, \ldots, i_d)$, where $i_k = 1, 2, \ldots, I_k$ for each $k = 1, 2, \ldots, d$. The mode-$(N, 1)$ product of two tensors $\mathcal{A} \in \mathbb{R}^{I_1 \times I_2 \times \cdots \times I_N}$ and $\mathcal{B} \in \mathbb{R}^{J_1 \times J_2 \times \cdots \times J_M}$ is defined as the tensor $C = \mathcal{A} \times_N^1 \mathcal{B}$, which results in a tensor $C \in \mathbb{R}^{I_1 \times I_2 \times \cdots \times I_{N-1} \times J_2 \times \cdots \times J_M}$. The elements of C are given by $c_{i_1, \ldots, i_{N-1}, j_2, \ldots, j_M} = \sum_{i_N=1}^{I_N} a_{i_1, \ldots, i_N} b_{j_1, \ldots, j_M}$ where $i_N = j_1$.

9.2.2 Intra-Source Transferable Feature Learning

For the TSFAN, the backbone network is inspired by the findings in [10–12], which highlight the varying robustness of different frequency bands to data segments across sessions. Specifically, we address the discrepancies across multiple frequency bands due to the inherent differences in the EEG signals' frequency components. Table 9.1 provides a detailed description of the backbone network for a single source domain.

Specifically, the temporal feature extractor $F_t(\cdot)$ employs multi-scale temporal kernels designed to adapt to different frequency bands. The kernel sizes are chosen in proportion to the EEG signal's sampling rate. Intuitively, the longer temporal kernels are capable of capturing long-term temporal dependencies in low-frequency representations, while shorter temporal kernels are suited for higher frequency components. Let the EEG input samples be denoted as $x \in \mathbb{R}^{c \times f \times t}$, where c represents the number of channels, f is the number of narrow frequency bands, and t is the length of each EEG sample. The temporal features are generated by concatenating the outputs of parallel multi-scale temporal kernels applied to the input data, followed by batch normalization, activation, and dropout. For simplicity, we define the temporal features as $\mathbf{Zt}_{sj} = F_t^j(X_{sj})$ and $\mathbf{Zt}_{tj} = F_t^j(X_{tj})$. The spatial-frequency feature extractor $F_{sf}(\cdot)$ is composed of spatial convolution layers and frequency-depthwise separable convolution layers. This design allows the network to learn both spatial and frequency-specific representations. The spatial-frequency features are defined as $\mathbf{Z}_{sj} = F_{sf}^j(\mathbf{Zt}_{sj})$ and $\mathbf{Z}_{tj} = F_{sf}^j(\mathbf{Zt}_{tj})$.

Next, the intra-source features $\mathbf{Z}_{sj} = \{\mathbf{z}_i^{sj}\}_{i=1}^{|\mathbf{Z}_{sj}|}$ are used to train the corresponding classifiers $C(\cdot)$, while the specific-source domain-invariant features $\mathbf{Z}_{tj} = \{\mathbf{z}_i^{tj}\}_{i=1}^{|\mathbf{Z}_{tj}|}$ from the target domain are employed for classifier alignment. These classifiers are trained by minimizing the cross-entropy loss, formulated as:

Table 9.1 Details of the backbone convolutional neural network

Block	Layer	Size	Option
Temporal Feature Extractor	Conv3D_1	(1, 1, t),	filter = 32
		(1, 1, t//2),	
		(1, 1, t//4)	
	BatchNorm		32
	Activation		Relu
	Dropout		0.25
Spatial-frequency Feature Extractor	Conv3D_2	(c, 1, 1)	filter = 64, group = 8
	BatchNorm		64
	Activation		Relu
	Dropout		0.25
	Conv3D_3	(1, 1, 4)	filter = 128, group = 8
	Conv3D_4	(1, 1, 1)	filter = 128
	BatchNorm		128
	Activation		Relu
	AveragePool3D	(1, 1, t)	
	Dropout		0.25

$$\mathcal{L}_{cls} = \sum_{j=1}^{N} \mathbb{E}_{x \sim X_{sj}} J(C_j(F_{sf}^j(F_t^j(x_i^{sj}))), y_i^{sj}), \quad (9.1)$$

where $J(\cdot, \cdot)$ denotes the cross-entropy loss function. For each sample in the target domain, predictions are made by N classifiers, one for each source domain. However, because of the inherent distributional differences between source and target domains, it is important to minimize the discrepancies across these classifiers. To achieve this, the discrepancy loss $\mathcal{L}_{disc}$ is introduced, which measures the absolute difference between the predictions of the classifiers for target domain data:

$$\mathcal{L}_{disc} = \frac{2}{N \times (N-1)} \sum_{j=1}^{N-1} \sum_{i=j+1}^{N} \mathbb{E}_{\mathbf{z} \sim \mathbf{Z}_t} \left[|C_i(\mathbf{z}_{ti}) - C_j(\mathbf{z}_{tj})| \right]. \quad (9.2)$$

In addition, the distributional discrepancy between each pair of source and target domains is quantified using the distribution distance loss $\mathcal{L}_{dist}$ as:

$$\mathcal{L}_{dist} = \frac{1}{N} \sum_{i=1}^{N} D(\mathbf{Z}_{ti}, \mathbf{Z}_{si}), \quad (9.3)$$

9.2 Tensorized Spatial-Frequency Attention Network

where D represents the discrepancy measure between the source and target domains. To align the feature distributions across domains, the Maximum Mean Discrepancy (MMD) method is employed, which is a well-established technique in deep domain adaptation. The overall objective of our TSFAN is to minimize the total loss function, which is expressed as:

$$\mathcal{L}_{total} = \mathcal{L}_{cls} + \lambda \mathcal{L}_{dist} + \gamma \mathcal{L}_{disc}, \quad (9.4)$$

where λ and γ are trade-off parameters that balance the contributions of the different loss terms.

9.2.3 Tensorized Spatial-Frequency Attention Mechanism

The instability of brainprint features across sessions is often caused by minor shifts in electrode placement and changes in human states. Previous research has indicated that distinct brain regions and specific frequency bands in EEG signals correspond to different human states. As a result, it is essential to extract consistent spatial-frequency features that remain invariant across various domains and perspectives. Building upon the concept of Tucker decomposition, which represents interactions between components through a core tensor, we introduce a tensorized spatial-frequency attention mechanism. This method captures the interactions between identity features from multiple domains by using a low-rank tensor, thereby emphasizing features that are invariant across domains. This approach effectively bypasses the need for aligning the distributions of source domain data. By leveraging multi-view domain-invariant features, the proposed mechanism reduces the impact of session instability by enabling interactions between features across different views, ensuring resilience against domain shifts among various sources. The schematic of the tensorized spatial-frequency attention mechanism is illustrated in Fig. 9.2. The temporal features from multiple source domains, represented in different colors, are fed into the spatial-frequency attention mechanism. This process begins with a nonlinear mapping performed by $\{F_a^i(\cdot)\}_{i=1}^N$ for each source domain. Then, the tensorized multi-view interaction is learned through the function $F_b(\cdot)$, which incorporates an interactive attention core tensor $\mathcal{U} \in \mathbb{R}^{r_1 \times r_2 \times \cdots \times r_{N+1}}$ and a set of factor matrices $\{\mathbf{W}_n \in \mathbb{R}^{r_n \times I_n}\}_{n=1}^{N+1}$. After the interaction is computed, the spatial-frequency attentions are reshaped and adapted to align with the input temporal features.

In particular, the temporal dimension of the temporal features $\mathbf{Zt}'_{sj}$ is initially reduced using a max-pooling layer, which compresses the global temporal information into a spatial-frequency descriptor $\mathbf{P}_{sj} \in \mathbb{R}^{c \times f}$. This is followed by two fully connected (FC) layers to perform a nonlinear transformation. The resulting spatial-frequency attention $\mathbf{Q}_{sj}$ is computed as follows:

$$\mathbf{Q}_{sj} = F_b^j \left(Relu \left(F_a^j \left(\mathbf{P}_{sj}; \mathbf{V_j} \right) \right); \mathbf{U_j} \right), \quad (9.5)$$

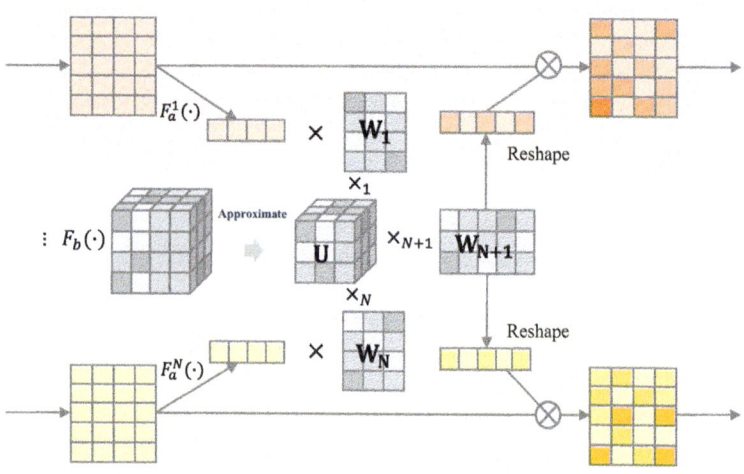

Fig. 9.2 Tensorized spatial-frequency attention mechanism

where $j \in [1, N]$, $F_a^j(\cdot)$ and $F_b^j(\cdot)$ represent two fully connected layers, with parameters $\mathbf{V_j}$, and $\mathbf{U_j}$ respectively, for the jth source domain.

While the previous mechanism enhances the intra-source transferable spatial-frequency features, it primarily focuses on refining features within each source domain. To enable the interaction of rich representations across multiple source views, the second FC layer's parameters, $\mathbf{U_j} \in \mathbb{R}^{c'f' \times cf}$, are tensorized into a single $(N+1)$-order tensor $\mathcal{W} \in \mathbb{R}^{c'f' \times c'f' \times \cdots \times cf}$, where $c' = c//3$, $f' = f//3$. Considering the challenge of high dimensionality, Tucker format [13] is further utilized to efficiently approximate the $\mathcal{W}$. This decomposition expresses $\mathcal{W}$ using a core tensor $\mathcal{U} \in \mathbb{R}^{r_1 \times r_2 \times \cdots \times r_{N+1}}$, along with factor matrices $\mathbf{W}_n \in \mathbb{R}^{r_n \times I_n}$ for $n = 1, \ldots, N+1$, where $I_1 = I_2 = \cdots = I_N = c'f'$ and $I_{N+1} = cf$. The factorization is given by:

$$\mathcal{W} = \mathcal{U} \times_1^2 \mathbf{W}_1 \times_2^2 \mathbf{W}_2 \times \cdots \times_{N+1}^2 \mathbf{W}_{N+1}, \tag{9.6}$$

where $[r_1, r_2, \ldots, r_{N+1}]$ represent the Tucker ranks. Notably, the Tucker ranks are typically much smaller than the original tensor dimensions. In this decomposition, the core tensor $\mathcal{U}$ captures interactions between components, enabling it to model the relationships across source views through its attention mechanism. Thus, the $\{F_b^i(\cdot)\}_{i=1}^N$ layers across the source domains can be unified in the following expression:

$$F_b(\mathbf{P}'_{sj}; \mathbf{U}, \mathbf{W}_1, \ldots \mathbf{W}_{N+1}) = \\ \mathcal{U} \times_1^2 \mathbf{W}_1 \cdots \times_i^2 (\mathbf{W}_i \times Fla(\mathbf{P}'_{sj})) \times \cdots \times_{N+1}^2 \mathbf{W}_{N+1}, \tag{9.7}$$

where $\mathbf{P}'_{sj} = F_a^j(\mathbf{P}_{sj})$, and $Fla(\cdot)$ denotes a flattening operation. The spatial-frequency attention $\mathbf{Q}_{sj}$ is then computed by applying Maxpooling to the output of $F_b(\cdot)$, excluding the $(N+1)^{th}$ dimensional. Finally, the temporal features of the source domain $\mathbf{Zt}_{sj}$ are adjusted by the attention mechanism as follows:

$$\mathbf{Zt}_{sj} = \mathbf{Zt}'_{sj} + \mathbf{Zt}'_{sj} \otimes Sigmoid(\mathbf{Q}_{sj}), \quad (9.8)$$

where $\otimes$ denotes the element-wise dot product. This tensorized spatial-frequency attention block operates as a residual module within the TSFAN architecture. A similar operation is applied to the temporal features of the target domain, $\mathbf{Zt}_{tj}$, as:

$$\mathbf{Zt}_{tj} = \mathbf{Zt}'_{tj} + \mathbf{Zt}'_{tj} \otimes Sigmoid(\mathbf{Q}_{tj}). \quad (9.9)$$

It is important to highlight that by incorporating the low-rank constraints in the interactive attention core tensor, the number of parameters for the TSFAN model grows linearly with the number of source domains, N.

9.3 Brain Fingerprint Identification with TSFAN

9.3.1 Data Pre-processing

Experiments are conducted on two multi-session EEG datasets, where the EEG signals are collected over multiple days across different sessions. All subjects are informed about the data acquisition process.

Dataset 1: 128-channels Multi-Task EEG Dataset (128-MTED) [6]. Power Spectral Density (PSD) describes how the signal's power is distributed across frequencies. Previous studies [2, 6] have shown that short-window estimated PSD features are effective for EEG-based biometric recognition. In this section, we follow the data setup from [6], extracting the 3–30 Hz PSD features for each EEG channel with a window size of 0.36 s and no overlap. The data is segmented into 15-second samples. To simulate a realistic identification scenario with a minimal number of channels, we use 9 channels from the standard 10–20 system, covering the Frontal, Central, Parietal, and Occipital regions, specifically Fz, F7, F8, C3, C4, P7, P8, O1, and O2. Given the varying number of EEG acquisitions for each subject in this dataset, two data splits are used to validate the proposed method: (1) Split-1: The last 60% of the sessions (rounded to the nearest integer) are used as labeled source domains for training, while the remaining sessions serve as the unlabeled target domain. (2) Split-2: The first 60% of the sessions are used as labeled source domains, with the remaining sessions serving as the unlabeled target domain.

Dataset 2: SJTU Emotion EEG Dataset IV (SEED-IV) [14]. Similar to Dataset 1, the 3–30 Hz PSD features are extracted from each channel with a window size of 0.5 s without overlap, and the data is segmented as 15 s samples. For validation, a leave-one-session-out strategy is applied, where each session is treated as the unlabeled target domain, while the remaining sessions are used as labeled source domains.

9.3.2 Baseline Methods

This section describes the baseline approach for comparison with our proposed TSFAN. The existing techniques can be classified into three categories: (1) Deep Learning without Domain Adaptation (DL w/o DA): This category represents deep learning approaches that do not include domain adaptation; (2) Source combine: In these methods, all source domains are merged into a single source. Notably, the same distribution distance loss used in TSFAN, specifically the Maximum Mean Discrepancy (MMD), is incorporated into the "DL w/o DA" models; (3) Multi-source: This category involves aligning the data distributions between each pair of source and target domains. The following models represent different approaches in these categories:

- **EEGNet** [15]: a CNN that employs one-dimensional convolutions to automatically extract temporal, spatial, and frequency features from EEG signals. It has been successfully applied to EEG-based biometric recognition [2, 6].
- **EEGNet-PSD** [6]: An adaptation of EEGNet designed to work with PSD features, this model has also been used for EEG-based biometric applications.
- **CNN-RNN** [2]: A hybrid model combining CNN and BiLSTM networks to capture the temporal dependencies of EEG data for identity feature extraction.
- **MEERNet** [8]: A multi-source domain adaptation method for EEG signals that focuses on extracting both domain-invariant and domain-specific features.
- **MTDANN** [16]: A multi-source domain adaptation approach for EEG signals, which integrates EEGNet with a domain discriminator network to learn invariant representations between source and target domains.

9.3.3 Implementation Details

All experiments were implemented in the PyTorch framework and trained in an end-to-end manner using stochastic gradient descent (SGD) with a momentum of 0.9 and a learning rate of 0.05. For the hyperparameters λ and γ in Eq. (9.4), we chose $\lambda = \gamma = 0.5$ for the 128-MTED dataset and set $\lambda = 1$, $\gamma = 0.25$ for the SEED-IV dataset. The tensor ranks for the Tucker network were configured to 2 and 5 for two validations of the 128-MTED dataset, and 2, 5, and 4 for three validations of the SEED-IV dataset. We also provide a detailed analysis of how the trade-off parameters and tensor ranks influence the performance of TSFAN. Model performance is evaluated using rank-1 classification accuracy (ACC) and equal error rate (EER) in closed-set verification, consistent with the approach in [6]. A higher value for ACC is considered better, as indicated by the upward arrow, whereas lower values for EER are desirable, as indicated by the downward arrow.

9.3.4 Performance Comparison of TSFAN with Baseline Methods

This section compares the performance of the proposed TSFAN with several baseline methods using the 128-MTED and SEED-IV datasets. The results, presented in Tables 9.2 and 9.3, clearly indicate that TSFAN outperforms all baseline models across all evaluation metrics. When compared to methods that combine multiple source domains, TSFAN demonstrates a significant improvement over the best-performing baseline, EEGNet-PSD, with an increase of 6.72% in ACC and a reduction of 0.87% in EER on the 128-MTED dataset. On the SEED-IV dataset, TSFAN achieves an impressive average accuracy of 94.10% and a low EER of 3.65%, outperforming most of the other methods tested. These results highlight the importance of extracting domain-invariant features from each source-target domain pair for robust performance. Furthermore, TSFAN shows a notable advantage over multi-source domain adaptation methods in terms of both average ACC and EER, highlighting the effectiveness of the interactions between multi-source views. The confusion matrices for two datasets, shown in Figs. 9.3 and 9.4, further confirm the model's ability to maintain stable and reliable recognition across different sessions for most subjects.

To evaluate the transferability of features learned by our TSFAN, we visualize the latent features captured by EEGNet-PSD (source combine), MTDANN, MEERNet,

Table 9.2 Performances of baselines and TSFAN on 128-MTED dataset with ACC (%) and EER (%)

Models		Split-1		Split-2		Average	
		ACC	EER	ACC	EER	ACC	EER
DL w/o DA	EEGNet	75.83	8.37	83.39	5.70	79.61	7.04
	CNN-RNN	76.36	6.99	79.94	4.95	78.15	5.97
	EEGNet-PSD	80.90	5.04	82.06	5.04	81.48	5.04
Source combine	EEGNet	74.09	8.13	83.89	4.54	78.99	6.34
	CNN-RNN	78.47	6.60	81.81	4.49	80.14	5.55
	EEGNet-PSD	82.22	4.56	82.68	4.59	82.45	4.58
Multi-source	MEERNet	76.11	6.25	81.37	4.55	78.74	5.40
	MTDANN	76.81	6.37	80.47	5.20	78.64	5.78
	TSFAN (Ours)	**87.78**	**4.49**	**92.35**	**2.94**	**90.07**	**3.71**

Table 9.3 Performances of baselines and TSFAN on SEED-IV dataset with ACC (%) and EER (%)

Models		→ Session 1		→ Session 2		→ Session 3		Average	
		ACC	EER	ACC	EER	ACC	EER	ACC	EER
DL w/o DA	EEGNet	51.90	19.96	71.70	12.88	67.61	12.47	63.74	15.10
	CNN-RNN	54.97	21.90	76.07	8.04	62.98	15.02	64.68	14.98
	EEGNet-PSD	58.54	17.25	72.49	8.45	68.12	17.59	66.39	14.43
Source combine	EEGNet	72.68	11.80	93.83	3.63	83.80	10.81	83.44	8.75
	CNN-RNN	72.63	14.12	89.57	2.99	75.95	9.09	79.38	8.73
	EEGNet-PSD	87.86	5.95	99.80	0.08	79.69	11.81	89.12	5.95
Multi-source	MEERNet	82.13	6.97	89.40	4.33	75.29	11.62	82.27	7.64
	MTDANN	55.83	18.06	73.41	10.42	63.92	15.94	64.38	14.81
	TSFAN (ours)	**89.33**	**5.29**	**99.95**	**0.02**	**93.04**	**5.63**	**94.10**	**3.65**

and TSFAN under the "Split-2" condition for the 128-MTED dataset and "Session 3" as the target domain for SEED-IV. As shown in Fig. 9.5, multi-source domain adaptation methods outperform the Source Combine method, particularly in mitigating domain shift effects between multiple sources. To assess the alignment of class distributions across domains, MMD is used to quantify the discrepancy between source and target domains. The average discrepancies for each class between the source and target domains are 2.17, 3.16, 3.95, and 1.36 for EEGNet-PSD (source combine), MEERNet, MTDANN, and TSFAN on the 128-MTED dataset, respectively. For the SEED-IV dataset, the class discrepancies are 3.12, 3.18, 3.52, and 2.52 for EEGNet-PSD (source combine), MEERNet, MTDANN, and TSFAN. A smaller discrepancy indicates better alignment and model performance. Notably, the latent feature distribution generated by TSFAN is more concentrated within each class, which can be attributed to the tensorized spatial-frequency attention mechanism. In contrast, baseline methods exhibit greater class independence and misalignment between source and target domains. These visualizations confirm that TSFAN effectively captures domain-invariant features, improving model discriminability without being hindered by distributional differences across domains.

Additionally, Fig. 9.6 illustrates the training data size for different subjects and the corresponding recognition accuracy on the 128-MTED dataset under the "Split-2" validation setting, considering the influence of training data imbalance.

To examine the impact of varying training sample sizes across subjects, we note that in multi-source domain adaptation, EEG data from multiple sessions of a subject are leveraged to learn feature spaces corresponding to distinct source domains. In cases where a subject only has one session of data available for training, that session is used to model the different source domain feature spaces. As observed, for subjects with fewer training samples (e.g., S29, S19, S05, etc.), the performance of comparison models experiences a significant decline, particularly when compared to subjects with larger sample sizes. In contrast, the proposed TSFAN exhibits a smaller performance gap across subjects, regardless of their sample sizes. This suggests that

9.3 Brain Fingerprint Identification with TSFAN

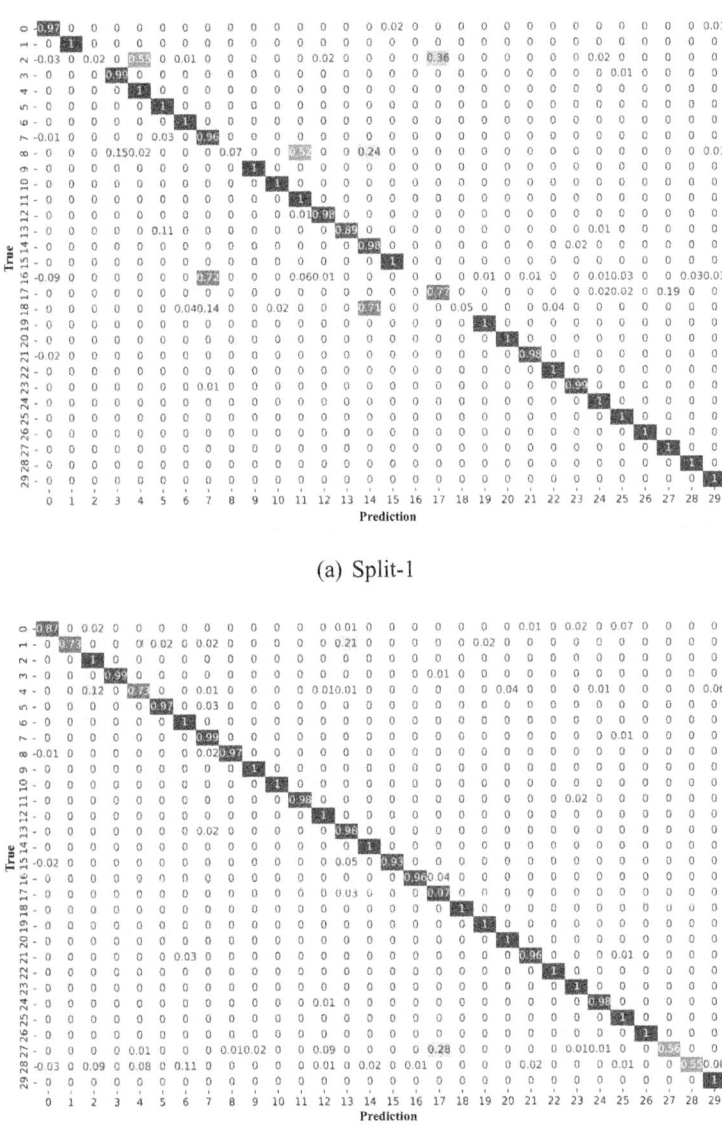

(a) Split-1

(b) Split-2

Fig. 9.3 Confusion matrix for TSFAN on the 128-MTED dataset

TSFAN is capable of more robustly extracting discriminative identity features from EEG data, effectively mitigating the negative effects of training sample imbalance. Interestingly, for some subjects with larger amounts of training data (e.g., S17, S01, S09), the performance of baseline methods is still lower than expected. This can be

Fig. 9.4 Confusion matrix for the performances of TSFAN on SEED-IV dataset

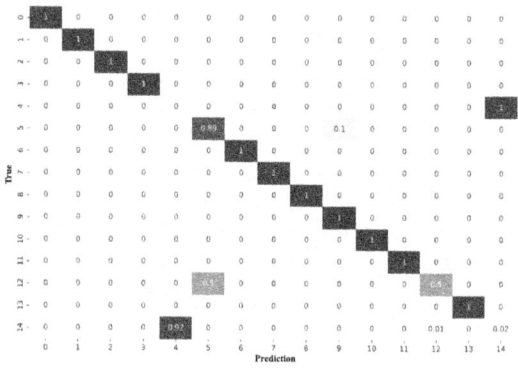

(a) Session 1

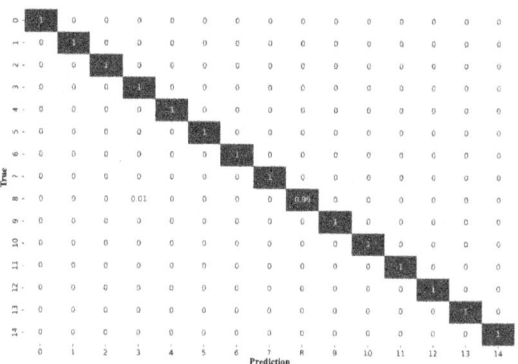

(b) Session 2

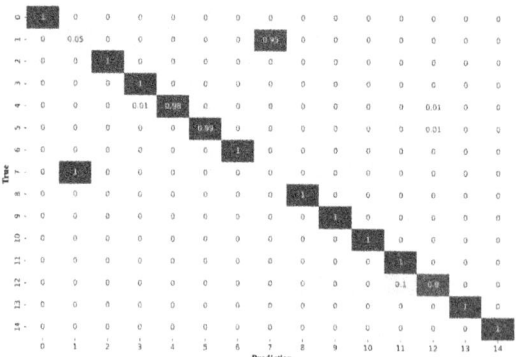

(c) Session 3

9.3 Brain Fingerprint Identification with TSFAN

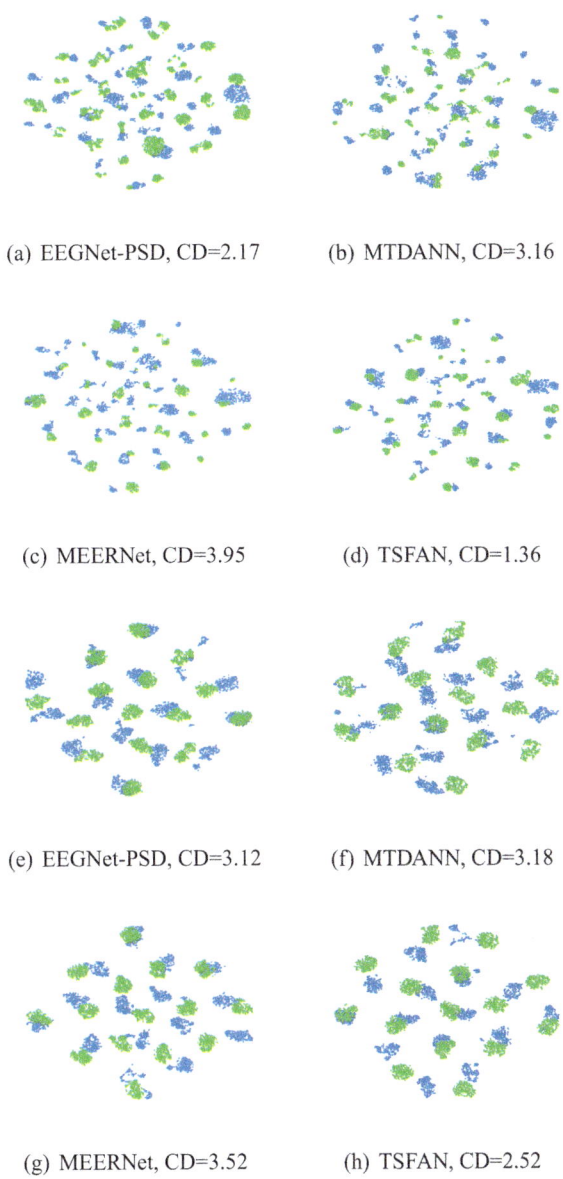

Fig. 9.5 t-SNE visualization of latent representations learned by EEGNet-PSD (Source combine), MTDANN (Multi-source), MEERNet (Multi-source), and TSFAN. The visualizations correspond to the "Split-2" setting for the 128-MTED dataset (**a–d**) and "Session 3" as the target domain for the SEED-IV dataset (**e–h**). In each plot, the class discrepancy between source and target domains is denoted as CD. Blue dots represent the first session as the source domain, while green dots indicate the target domain

(a) EEGNet-PSD, CD=2.17
(b) MTDANN, CD=3.16
(c) MEERNet, CD=3.95
(d) TSFAN, CD=1.36
(e) EEGNet-PSD, CD=3.12
(f) MTDANN, CD=3.18
(g) MEERNet, CD=3.52
(h) TSFAN, CD=2.52

attributed to the substantial distributional differences between EEG data acquired across sessions, which often hampers performance in cross-session recognition scenarios. In contrast, TSFAN addresses this challenge by utilizing multi-source domain adaptation, allowing for more consistent and reliable recognition performance, even when data distributions across sessions vary significantly.

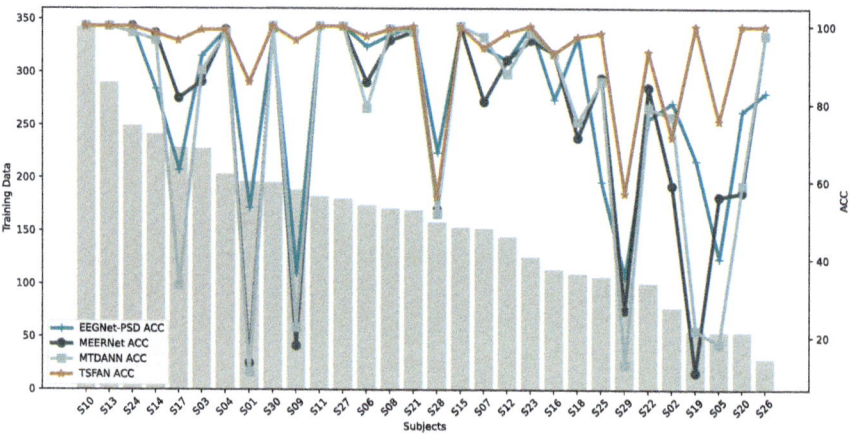

Fig. 9.6 Size of training data and performances of baselines and TSFAN on 128-MTED dataset for each subject

Table 9.4 Complexity comparison of baselines and TSFAN on 128-MTED dataset

Models	Parameters	Training time (s/epoch)	Testing time (s/epoch)
EEGNet	322,622	5.14	32.68
CNN-RNN	17,766,718	0.21	9.68
EEGNet-PSD	40,062	0.21	9.74
MEERNet	118,714	0.66	9.89
MTDANN	170,877	0.67	9.93
TSFAN (Ours)	788,227	0.88	9.93

Additionally, the complexity of the models is assessed by analyzing the number of model parameters, as well as the training and testing times for each batch and all test samples, respectively. The results are summarized in Table 9.4, where the tensor rank of TSFAN is set to 5. Notably, each source feature space in TSFAN contains 261,274 parameters, while the tensorized interactive attention mechanism contributes an additional 4,405 parameters. Most of the parameters in TSFAN are derived from the multi-scale convolutional layers. However, when comparing the training and testing times, TSFAN shows performance on par with the baseline models. This efficiency is largely due to the application of low-rank Tucker decomposition, which ensures that the number of parameters and computational complexity increase linearly with the number of source domains, rather than exponentially. This characteristic allows TSFAN to maintain computational efficiency while leveraging the rich feature interactions across multiple source domains.

9.3.5 Effect of Tensorized Spatial-Frequency Attention Mechanism

In this section, we isolate the independent contributions of the proposed TSFAN by defining several variants for verification, which are outlined as follows:

- **TSFAN w/o attention**: A version of TSFAN that excludes the Tensorized spatial-frequency attention block, removing the attention mechanism entirely.
- **TSFAN w/o interaction**: TSFAN with the spatial-frequency attention applied independently to each specific-source domain, without the tensorized interaction mechanism.
- **TSFAN w/o Tucker**: TSFAN that employs the tensorized spatial-frequency attention but without the low-rank Tucker decomposition for efficient tensor representation.

These variants are designed to investigate the individual contributions of each component in the proposed TSFAN framework, helping to better understand the impact of the attention mechanism, multi-source interaction, and Tucker decomposition on overall performance.

The results of the ablation studies are summarized in Tables 9.5 and 9.6. Interestingly, the EER of the variant "TSFAN w/o attention" is comparable to that of other multi-source domain adaptation methods. This suggests that our backbone model effectively captures intra-source transferable features, which remain crucial even without the attention mechanism. A comparison between TSFAN and "TSFAN w/o interaction" reveals that TSFAN outperforms "TSFAN w/o interaction" by 2.16% in average ACC and achieves a 0.6% lower EER on the 128-MTED dataset. For the SEED-IV dataset, TSFAN boosts ACC by 9.56% and reduces EER by 4.78%. These results demonstrate the critical importance of inter-source interactions in enhancing domain-invariant feature extraction. When compared to "TSFAN w/o Tucker", our TSFAN shows a significant performance advantage, reinforcing the notion that tensorizing the source-specific attention without Tucker decomposition is highly vulnerable to the curse of dimensionality. The low-rank structure provided by the Tucker decomposition effectively mitigates this issue, validating the strength and efficiency of the proposed method.

9.3.6 Effect of Tensor Ranks for Tucker Network

Since tensor ranks are pivotal to the effectiveness of our tensorized spatial-frequency attention mechanism, we investigate the impact of different tensor ranks on the performance of TSFAN. The tensor ranks, denoted as $r_1 = r_2 = \cdots = r_{N+1}$, are varied from 2 to 7. The results are shown in Fig. 9.7. The experimental findings suggest that the model tends to achieve optimal performance when the tensor rank is less than or equal to 5. Specifically, for the 128-MTED dataset, the best performance is observed

Table 9.5 Performances of ablation methods on 128-MTED dataset with ACC (%) and EER (%)

Models	Split-1		Split-2		Average	
	ACC	EER	ACC	EER	ACC	EER
TSFAN w/o attention	84.16	5.25	88.25	5.15	86.21	5.20
TSFAN w/o interaction	85.45	5.20	90.38	3.42	87.91	4.31
TSFAN w/o tucker	85.69	5.76	91.44	3.21	88.56	4.48
TSFAN	**87.78**	**4.49**	**92.35**	**2.94**	**90.07**	**3.71**

Table 9.6 Performances of ablation methods on SEED-IV dataset with ACC (%) and EER (%)

Models	→ Session 1		→ Session 2		→ Session 3		Average	
	ACC	EER	ACC	EER	ACC	EER	ACC	EER
TSFAN w/o attention	75.43	13.99	95.49	3.66	88.92	6.69	86.62	8.11
TSFAN w/o interaction	80.48	10.93	94.41	2.16	78.72	12.19	84.54	8.43
TSFAN w/o tucker	81.88	10.55	95.50	3.25	88.71	7.48	88.70	7.09
TSFAN	**89.33**	**5.29**	**99.95**	**0.02**	**93.04**	**5.63**	**94.10**	**3.65**

for ranks of 2 and 5 under both "Split-1" and "Split-2" settings. Similarly, for the SEED-IV dataset, the rank values of 5, 2, and 4 provide the best performance. These findings highlight the importance of choosing an appropriate tensor rank to balance model complexity and performance.

The findings indicate that using a lower tensor rank enhances the model's ability to capture the underlying patterns and features within the given dataset and experimental setup. Moreover, these results demonstrate that TSFAN effectively exploits the sparse correlations across multiple source domains, leading to more stable feature extraction for cross-session EEG-based identity recognition. By leveraging the low-rank representation of the Tucker Network, the model's parameter count scales linearly with the number of source domains (N), which significantly reduces both computational demands and storage requirements.

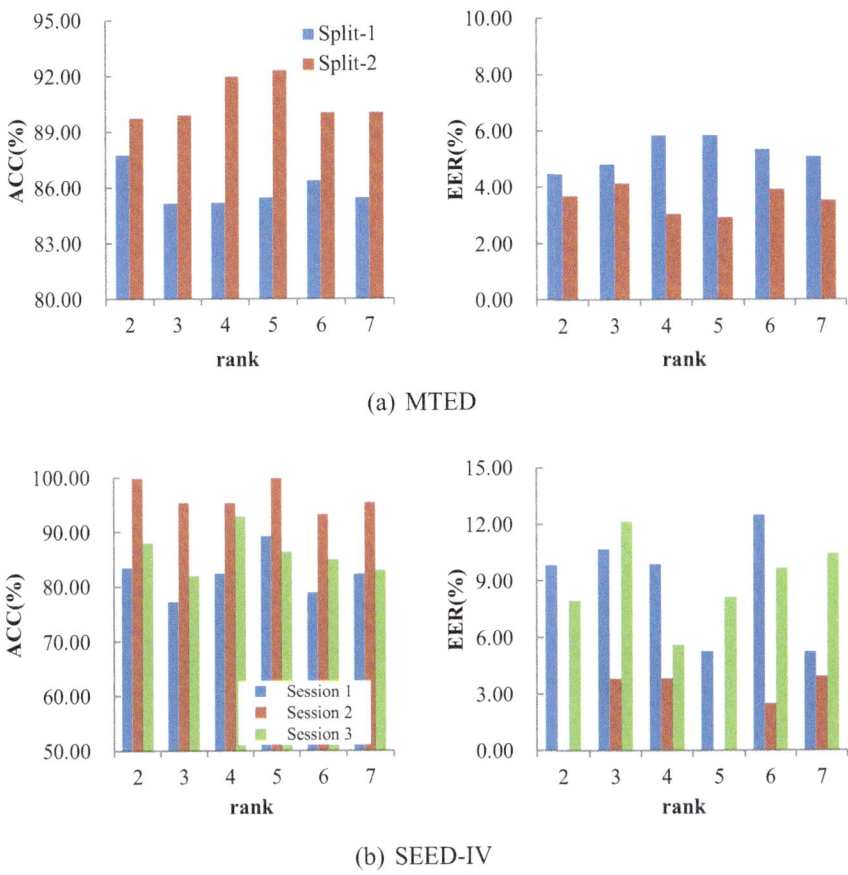

Fig. 9.7 The effect of tensor ranks in the Tucker network on performance for both the 128-MTED and SEED-IV datasets

9.3.7 Effect of Trade-Off Parameter λ and γ

In this section, the trade-off parameters λ and γ are analyzed. The parameter λ controls the weight of the distance loss between each pair of source and target domains, while γ is associated with the discrepancy loss to ensure alignment of the classifiers. Figure 9.8 illustrates the performance for various values of these parameters. Specifically, λ is set to 0.5, 1, and γ is tested with values in 0, 0.1, 0.25, 0.5, 1. The results show that the model performs better when $\lambda = 1$ compared to $\lambda = 0.5$. Furthermore, the performance exhibits a bell-shaped curve as γ increases, indicating an optimal point at higher values of γ. These findings suggest that aligning data distributions pairwise between source and target domains enhances the extraction of domain-invariant features. However, larger discrepancies across multiple source domains can hinder the stability of the features. Thus, effectively capturing inter-

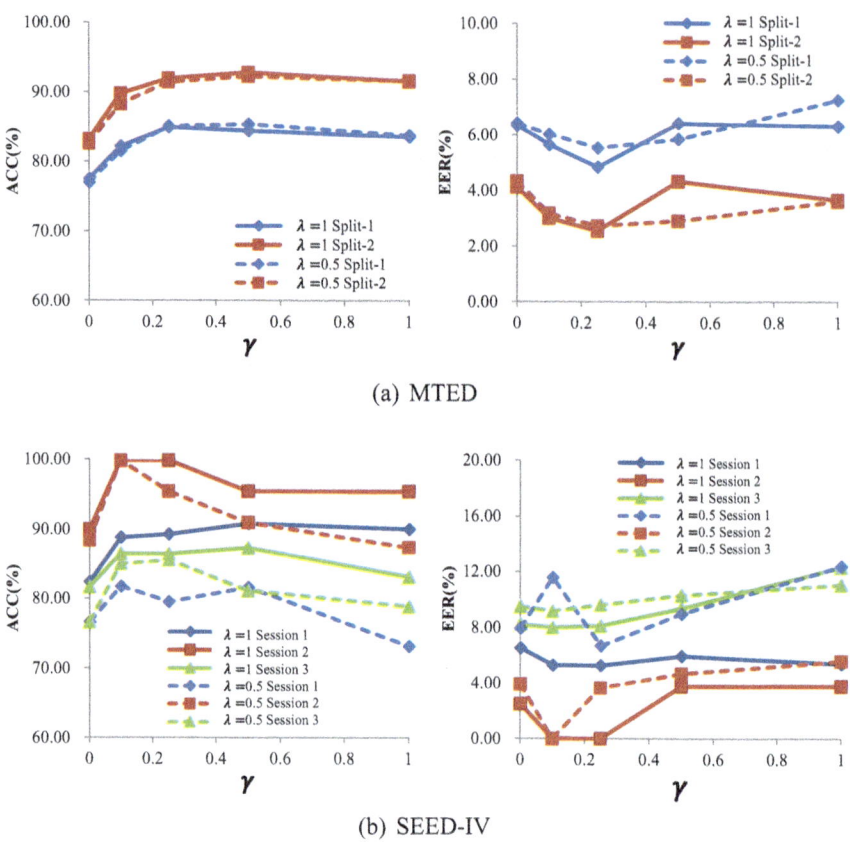

Fig. 9.8 The effect of the trade-off parameters λ and γ on the performance of TSFAN using the 128-MTED and SEED-IV datasets

actions between domain-invariant features from different source domains using a tensorized attention mechanism proves to be vital for optimal performance.

9.3.8 Effect of Electrodes for EEG-Based Biometric Recognition

In this section, the impact of electrode selection on EEG-based biometric recognition is analyzed. In practical applications, reducing the number of electrodes used is crucial for efficient performance, while still maintaining recognition accuracy. The goal is to identify the fewest electrodes that can reliably capture identity-related features in EEG data.

9.3 Brain Fingerprint Identification with TSFAN

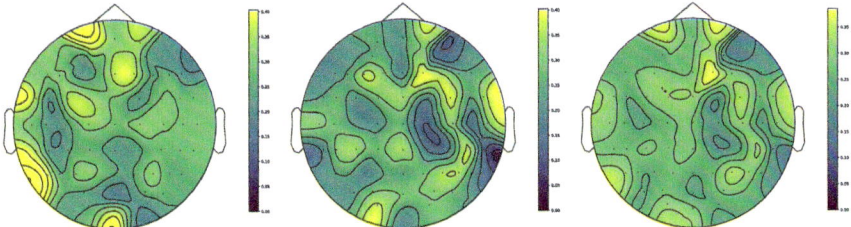

Fig. 9.9 The topography of electrode weight

To explore the importance of each electrode, Gradient-weighted Class Activation Mapping (Grad-CAM) [17] is applied to the EEG data. Grad-CAM highlights regions of the brain that are most relevant for identity prediction, allowing us to visualize the importance of different electrodes. For this experiment, the datasets are split chronologically, with earlier sessions used for training and later sessions for testing. The topographic maps in Fig. 9.9 depict the importance of electrodes for SEED-IV datasets, considering the temporality of the data. For SEED-IV dataset, the first two topographies are specific-source views, with the last one representing the target view. The results show that for the task-specific EEG dataset, the electrodes remain spread across regions of the brain, task-related electrodes play a more prominent role in identity recognition.

Based on these findings, the number of electrodes used in the experiments is reduced for both datasets. For 128-MTED, 3, 5, 9, and 19 electrodes are selected from the standard 10–20 system, and the reference electrode Cz is excluded. For SEED-IV, 3, 5, 20 electrodes from the 10–20 system, as well as 23 electrodes including task-related regions, are selected. These subsets are compared against the full set of electrodes for both datasets. Figure 9.10 illustrates the electrode selection and their corresponding positions. Tables 9.7 and 9.8 present the experimental outcomes for different electrode configurations, with tensor ranks set to 2. The results show that using 19 electrodes for 128-MTED and 20/23 electrodes for SEED-IV achieves performance comparable to the full electrode setup, with no significant differences in recognition accuracy. Notably, when "Session 2" and "Session 3" are used as the target domains, the performance with 20 electrodes is similar to that with 23 electrodes. Additionally, as highlighted in Fig. 9.10b and Table 9.8b, electrodes located in the frontal lobe and frontal pole lobe play a more significant role in identity recognition, particularly on the emotion-specific task dataset. This could be due to the involvement of these regions in the processing of mental tasks, where there is variability in how subjects process the same task, thus affecting the recognition of identity. Furthermore, Fig. 9.11 shows the performance of the 20 and 23-electrode configurations on SEED-IV for various tensor ranks. Interestingly, the performance with 20 electrodes is more stable. When 23 electrodes, which include task-related regions, are used, the best performance is achieved with a lower tensor rank of 2. However, as the tensor rank increases, performance starts to degrade. This suggests that the variability in mental tasks across sessions may be responsible for the performance

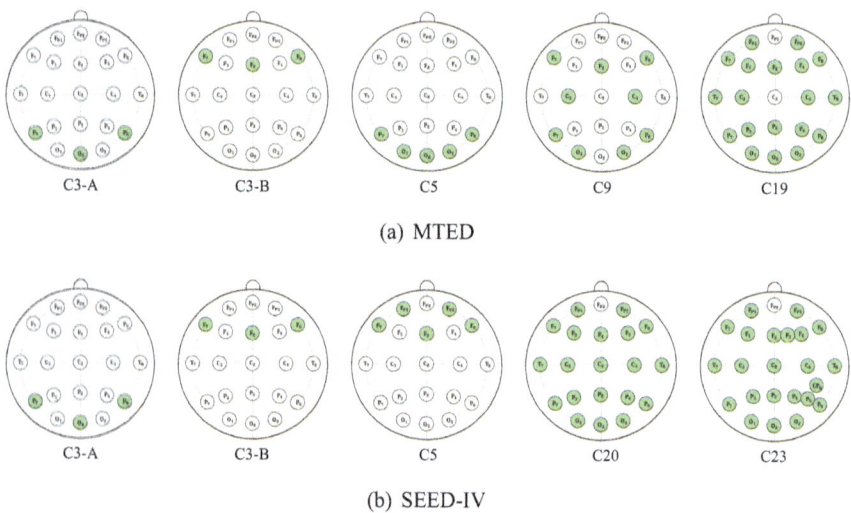

Fig. 9.10 The detailed electrode selection and position for 128-MTED and SEED-IV

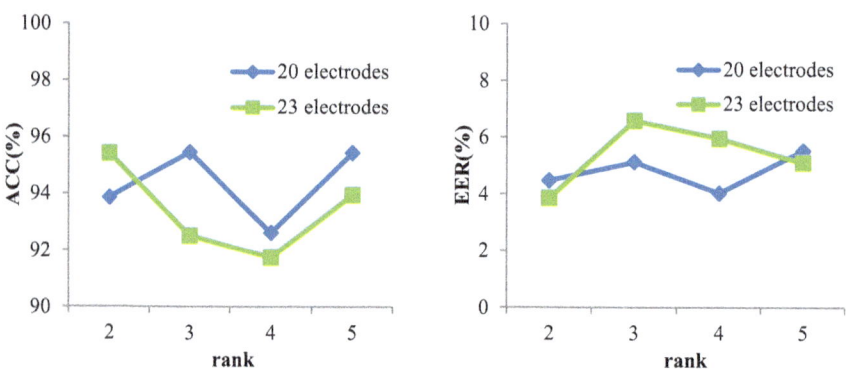

Fig. 9.11 The results of various tensor ranks for 20 electrodes and 23 electrodes on SEED-IV

Table 9.7 Performances of various electrode selection on 128-MTED with ACC (%) and EER (%)

Channel	C3-A	C3-B	C5	C9	C19	C128
ACC	67.14	69.35	69.35	92.35	96.14	**96.42**
EER	12.27	12.65	12.65	2.94	**1.03**	1.32
p-valve	***	***	***	***	–	–

drop, as mental task processing is more susceptible to changes in the subject's state and may affect the consistency of the identity-related information extracted.

9.4 Conclusion 157

Table 9.8 Performances of various electrode selection on SEED-IV with ACC (%) and EER (%)

Channel	C3-A	C3-B	C5	C20	C23	C62
ACC	43.45	57.04	77.41	86.51	**86.84**	86.53
EER	22.73	21.84	12.90	12.01	11.55	**8.17**
p-valve	***	***	***	–	–	–

9.4 Conclusion

In this chapter, we introduce TSFAN as a novel approach for brain fingerprint identification, addressing critical challenges in the stability and transferability of identity features across sessions. Unlike most existing methods, which fail to account for the complexities of inter-session interactions or the distribution shifts that arise across multiple sessions, TSFAN effectively captures stable identity-related EEG features. The key innovation of TSFAN lies in its use of tensor-based attention, where domain-specific attention is tensorized in a low-rank Tucker format. This design allows TSFAN to model the intricate relationships between domain-invariant temporal representations, enabling robust recognition despite the variability introduced by session shifts.

A notable advantage of TSFAN is its flexibility in handling an arbitrary number of sessions, thanks to the low-rank tensor representation. This scalability ensures that TSFAN can be applied to a wide range of real-world EEG-based biometric applications with varying amounts of data from different sessions. Additionally, our electrode selection analysis highlights that brainprint features are distributed across multiple brain regions, with the 20 electrodes chosen based on the standard 10–20 system being sufficient for stable identity feature extraction.

Despite these strengths, there are inherent limitations. The reliance on a tensor-based structure, while effective in reducing computational complexity, may still struggle with extreme variations in brain activity across subjects or sessions, particularly when task-related features dominate. Furthermore, while TSFAN has demonstrated flexibility in session-based scenarios, its performance could be influenced by the quality and consistency of the EEG data, especially in environments with high noise or subject-specific variations.

Overall, TSFAN represents a promising direction for advancing EEG-based biometric recognition, providing a robust, scalable solution for cross-session identity verification. However, further exploration is needed to refine its handling of task-specific and session-dependent variations in EEG data.

References

1. Zhang S, Sun L, Mao X, Hu C, Liu P (2021) Review on EEG-based authentication technology. Comput Intell Neurosci
2. Maiorana E (2020) Deep learning for EEG-based biometric recognition. Neurocomputing 410:374–386
3. Jin X, Tang J, Kong X, Peng Y, Cao J, Zhao Q, Kong W (2020) CTNN: a convolutional tensor-train neural network for multi-task brainprint recognition. IEEE Trans Neural Syst Rehabil Eng 29:103–112
4. Wilaiprasitporn T, Ditthapron A, Matchaparn K, Tongbuasirilai T, Banluesombatkul N, Chuangsuwanich E (2019) Affective EEG-based person identification using the deep learning approach. IEEE Trans Cogn Dev Syst 12(3):486–496
5. Wang M, El-Fiqi H, Hu J, Abbass HA (2019) Convolutional neural networks using dynamic functional connectivity for EEG-based person identification in diverse human states. IEEE Trans Inf Forensics Secur 14(12):3259–3272
6. Kumar MG, Narayanan S, Sur M, Murthy HA (2021) Evidence of task-independent person-specific signatures in EEG using subspace techniques. IEEE Trans Inf Forensics Secur 16:2856–2871
7. Liu D, Zhang J, Hanrui W, Liu S, Long J (2023) Multi-source transfer learning for EEG classification based on domain adversarial neural network. IEEE Trans Neural Syst Rehabil Eng 31:218–228
8. Chen H, Li Z, Jin M, Li J (2021) Meernet: multi-source EEG-based emotion recognition network for generalization across subjects and sessions. In: 2021 43rd annual international conference of the IEEE engineering in medicine & biology society (EMBC). IEEE, pp 6094–6097
9. Ding Z, Shao M, Fu Y (2018) Robust multi-view representation: a unified perspective from multi-view learning to domain adaption. IJCAI
10. da Silva Castanheira J, Orozco Perez HD, Misic B, Baillet S (2021) Brief segments of neurophysiological activity enable individual differentiation. Nat Commun 12(1):1–11
11. Zheng W-L, Bao-Liang L (2015) Investigating critical frequency bands and channels for EEG-based emotion recognition with deep neural networks. IEEE Trans Auton Ment Dev 7(3):162–175
12. Peng Y, Qin F, Kong W, Ge Y, Nie F, Cichocki A (2022) GFIL: a unified framework for the importance analysis of features, frequency bands, and channels in EEG-based emotion recognition. IEEE Trans Cogn Dev Syst 14(3):935–947
13. Kolda TG, Bader BW (2009) Tensor decompositions and applications. SIAM Rev 51(3):455–500
14. Zheng W-L, Liu W, Yifei L, Bao-Liang L, Cichocki A (2018) Emotionmeter: a multimodal framework for recognizing human emotions. IEEE Trans Cybern 49(3):1110–1122
15. Lawhern VJ, Solon AJ, Waytowich NR, Gordon SM, Hung CP, Lance BJ (2018) Eegnet: a compact convolutional neural network for EEG-based brain–computer interfaces. J Neural Eng 15(5):056013
16. Liu D, Zhang J, Wu H, Liu S, Long J (2022) Multi-source transfer learning for EEG classification based on domain adversarial neural network. IEEE Trans Neural Syst Rehabil Eng
17. Selvaraju RR, Cogswell M, Das A, Vedantam R, Parikh D, Batra D (2017) Grad-cam: visual explanations from deep networks via gradient-based localization. In: Proceedings of the IEEE international conference on computer vision, pp 618–626

Open Access This chapter is licensed under the terms of the Creative Commons Attribution-NonCommercial-NoDerivatives 4.0 International License (http://creativecommons.org/licenses/by-nc-nd/4.0/), which permits any noncommercial use, sharing, distribution and reproduction in any medium or format, as long as you give appropriate credit to the original author(s) and the source, provide a link to the Creative Commons license and indicate if you modified the licensed material. You do not have permission under this license to share adapted material derived from this chapter or parts of it.

The images or other third party material in this chapter are included in the chapter's Creative Commons license, unless indicated otherwise in a credit line to the material. If material is not included in the chapter's Creative Commons license and your intended use is not permitted by statutory regulation or exceeds the permitted use, you will need to obtain permission directly from the copyright holder.

Chapter 10
Cross-Task and Cross-Session Brain Fingerprint Identification with Disentangled Adversarial Generalization Network

Abstract EEG-based brainprint technology, capitalizing on the inherent invisibility of EEG signals, has garnered significant attention as a promising biometric approach capable of meeting the stringent security requirements of various high-stakes applications. Traditional studies in this domain have largely focused on identity recognition based on single mental tasks, yet such tasks often introduce interference from spontaneous brain activity, which in turn causes spurious correlations between identity and task-related information. This issue is further compounded by the inherent temporal variability of EEG signals, which leads to significant data distribution differences across sessions. Consequently, existing models frequently fail to generalize unseen data from different sessions or tasks, severely limiting the practicality and scalability of brain fingerprint identification systems in real-world scenarios. In this chapter, we propose the Disentangled Adversarial Generalization Network (DAGN), a novel deep learning framework aimed at achieving stable and robust task-independent brain fingerprint identification across sessions. The core innovation of the DAGN lies in its ability to disentangle identity-relevant and task-relevant features through a decorrelation mechanism. This process effectively eliminates the spurious correlations between the two, ensuring that identity information is isolated from task-induced variability. To further enhance the generalization ability of the model, we introduce an adversarial self-challenging strategy that penalizes the activation of task-related features, forcing the network to focus exclusively on identity-specific information. This approach significantly improves the robustness of the learned features, enabling the DAGN to maintain high performance when applied to unseen data across different tasks and sessions. Extensive experiments, conducted on representative multi-task benchmarks with challenging leave-one-task-out and leave-one-session-out cross-validation protocols, demonstrate the superiority of our approach over state-of-the-art methods in terms of generalization performance.

10.1 Introduction

Brain fingerprint identification has emerged as a promising biometric technology due to its foundation in the unique brain activity patterns of each individual, which are inherently linked to personal memories, experiences, and knowledge. This method is dependent on living subjects and is extremely difficult to replicate, providing a high level of security. Existing brain fingerprint identification paradigms exploit various EEG elicitation techniques, such as resting-state potentials, motor imagery (MI), and steady-state visual-evoked potentials (SSVEP). However, these paradigms typically require the subject's full cooperation during signal acquisition and assume that the subject does not have any underlying physiological deficits, thus limiting their applicability in real-world settings. To overcome these limitations, there is a critical need to explore brainprint features across a range of mental tasks, thereby enabling more robust and practical biometric systems.

In recent years, deep learning has garnered significant attention for its ability to decode task-independent brainprint features across multiple tasks [1–4]. This is largely due to deep learning's capacity to automatically extract relevant features and uncover latent dependencies within the data. Despite its promise, most existing studies have validated their models using the same mental state EEG data that was employed for training, which is not reflective of real-world conditions. Specifically, these studies often fail to address the more challenging cross-task scenario, where models must generalize to unseen tasks in a leave-one-task-out validation setup. This gap in the validation process limits the true practical utility of these approaches, highlighting the need for methods that can handle such cross-task variability.

An inherent characteristic of the nervous system is its continuous spontaneous variability, which has been shown to be relevant to identity-differentiating information and to exert variable effects on cognitive functions [5–7]. This variability implies that identity-relevant discriminative components and task-relevant components of EEG signals may be correlated, complicating the process of extracting generalizable features across different cognitive states. The overlap between these components makes it challenging to distinguish between identity-specific information and task-specific information, especially when transitioning from task-specific states to more generalized human states. Additionally, the temporal variability of EEG signals, which arises from fluctuations in individual states and external interference from devices, leads to significant distributional differences in EEG data collected across time. Deep learning methods, which tend to activate dominant features that are strongly correlated with labels, can perform well when the training and testing data come from similar distributions. However, this performance deteriorates when data from different distributions are encountered. The root cause lies in the fact that, in such cases, the learned features are often overfitted to local, dominant components, rather than capturing the global relevance of identity features [8, 9]. Generalizable features should ideally focus on capturing the global identity information, rather than merely relying on the local task-dependent components, which limits the applicability of these methods to real-world scenarios where variability is inherent and unavoidable.

To overcome the challenges of disentangling identity and task-related features and addressing temporal variability in EEG signals, this chapter proposes the Disentangled Adversarial Generalization Network (DAGN). For the first challenge, a nonlinear feature decorrelation technique based on Random Fourier Features(RFF) [10] is employed. This approach effectively isolates identity-specific information from task-dependent components, reducing the unintended correlations that hinder feature generalization. For the second challenge, the model incorporates an attention mechanism alongside an adversarial self-challenging strategy. This combination suppresses dominant task-related features in the training domain and encourages the extraction of global identity-relevant features, supported by an auxiliary classifier to enhance prediction accuracy in unseen conditions. The main contributions of this chapter are as follows:

- Introduction of a Disentangled Representation Learning Framework: This chapter presents a novel framework tailored for task-agnostic brain fingerprint identification. It addresses the limitations of traditional methods by resolving spurious associations between identity and task features, thereby enabling improved cross-task generalization.
- Design of an Adversarial Self-Challenging Mechanism: An innovative adversarial strategy is developed to push the model beyond reliance on dominant features. By forcing the exploration of less obvious but significant identity-relevant features, this mechanism ensures robust feature extraction across diverse sessions.
- Comprehensive Experimental Evaluation: Rigorous experiments are conducted using two widely recognized multi-task EEG datasets. The results, validated through challenging cross-task and cross-session protocols, underscore the superior generalization and adaptability of the DAGN compared to existing state of the-art approaches.
- Exploration of Task-Independent Brainprint Properties: This work investigates the characteristics of task-independent brain fingerprint identification in detail, analyzing the influence of factors such as signal length, frequency bands, and the number of EEG channels on performance stability.

10.2 Disentangled Adversarial Generalization Network

In this section, we introduce the DAGN, a novel framework specifically designed to address the challenges of cross-session and task-independent brain fingerprint identification. The proposed DAGN architecture is structured around three core components: (1) Primary Feature Extraction: A backbone network is employed to extract fundamental features from raw EEG signals, providing a robust foundation for subsequent processing stages; (2) Feature Disentanglement: A mechanism is implemented to disentangle identity-relevant features from task-related ones, thereby reducing the interference of task-specific information on identity representation. This disentanglement step is critical for ensuring that the extracted identity features remain invariant across diverse tasks; (3) Adversarial Self-Challenging with Attention: An attention mechanism, integrated with an adversarial self-challenging strategy, is applied to refine the generalization ability of identity-relevant features. This approach encourages the model to focus on global identity-specific information, enhancing classification performance in unseen conditions. A comprehensive overview of the proposed method is illustrated in Fig. 10.1, which highlights the interaction between these components and their contribution to the overall architecture. The backbone feature generators, G_{ID} and G_T, transform the input power spectral density (PSD) features into identity-relevant features f_{ID} and task-relevant features f_T, respectively. Linear and nonlinear dependencies between f_{ID} and f_T are eliminated using the RFF technique. Simultaneously, an attention mechanism is adversarially optimized through the identity classifier C_{ID} and an auxiliary identify classifier C'_{ID}. This adversarial training encourages the model to activate not only the dominant identity-relevant features but also the remaining label-related features, enabling the extraction of stable task-independent identity features across tasks and sessions.

This chapter formalizes the brain fingerprint identification task using a training dataset denoted as $\mathcal{D} = \{(X_{sj}, Y_{sj}, C_{sj})\}_{j=1}^{N}$, where $X = \{x_i\}_{i=1}^{|X_{sj}|}$ represents the input features derived from the PSD of EEG signals, $Y = \{y_i\}_{i=1}^{|Y_{sj}|}$ corresponds to the identity labels associated with each sample, and $C = \{c_i\}_{i=1}^{|C_{sj}|}$ denotes the labels of the mental tasks performed during data acquisition. The parameter N specifies the total number of training sessions included in the dataset. The following sections provide a detailed explanation of each component within the proposed framework, elucidating how these elements interact to enable robust and task-independent brain fingerprint identification.

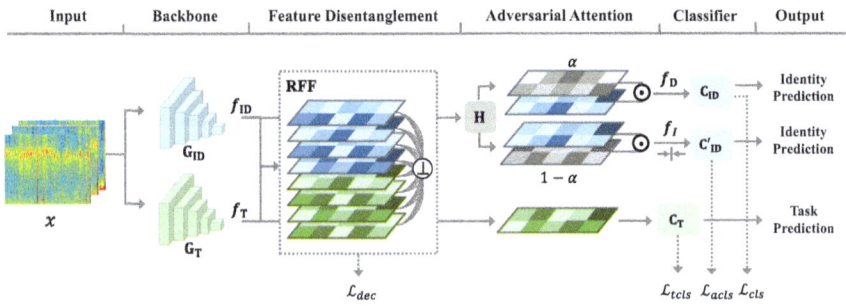

Fig. 10.1 The overall flowchart of the proposed DAGN framework

10.2.1 Backbone Convolutional Neural Network

In this section, a band-specific convolutional neural network (CNN) is employed as the backbone for extracting primary identity-relevant and task-relevant features. The detailed architecture of the backbone network is outlined in Table 10.1. The PSD, which describes the signal's intensity distribution in the frequency domain, has been shown in prior studies [1, 4] to be effective for brain fingerprint identification when estimated using short-time windows. Following this insight, we use 3D PSD features, estimated via the short-time Fourier transform (STFT), as input for individual identification. The input tensor to the backbone is defined as $x \in \mathbb{R}^{c \times s \times t}$, where c, s, and t represent the dimensions corresponding to channels, frequencies, and temporal samples, respectively. Given the multirhythmic nature of EEG signals and the variability across different frequency bands, previous studies [11, 12] have addressed these band-specific discrepancies to enhance feature extraction. Moreover, [13] demonstrated that multi-scale CNN architectures significantly improve EEG decoding performance by capturing complex temporal and spatial patterns. Drawing inspiration from these findings, the backbone network in this study is designed with specialized blocks to efficiently extract features from different dimensions: (1) "Block$_T$": This block is designed to capture temporal features within different frequency bands. It employs three parallel temporal-domain convolutional layers, followed by an average pooling layer, this structure enables the network to focus on temporal variations across multiple rhythms; (2) "Block$_S$": This block processes frequency-domain features using convolutional layers followed by average pooling. By focusing on temporal-frequency interactions, it enhances the extraction of frequency-dependent features; (3) "Block$_C$": The final block concatenates features

Table 10.1 Architecture of backbone convolutional neural network

Block	Layer	Size			Option
Block$_T$	Conv3D	(1, 1, t//2)	(1, 1, t//8)	(1, 1, t//4)	Filter = 32
	AveragePool3D	(1, 1, 5)			
	BatchNorm				
	Activation				Relu
Block$_S$	Conv3D	(1, s_1//2, 1)	(1, s_2//2, 11)	(1, s//2, 1)	Filter = 64
	Conv3D	(1, s_1//2, 1)	(1, s_2//2, 11)	(1, s//2, 1)	Filter = 128
	AveragePool3D	(1, 1, 9)	(1, 1, 12)	(1, 1, 5)	
	BatchNorm				
	Activation				Relu
	Dropout				0.25
Block$_C$	Conv3D	(c, 1, 3)			Filter = 128
	BatchNorm				
	Activation				Relu

s_1 and s_2 denote the dimensions of the frequency domain corresponding to low-frequency and high-frequency data, respectively from multiple frequency bands and applies a 3D convolutional layer to capture spatial features. This block ensures that the extracted features retain spatial coherence across EEG channels.

As illustrated in Fig. 10.1, the backbone network is employed to extract two types of features: identity-relevant feature f_{ID} and task-relevant feature f_T. Both feature types are derived using two CNNs with identical architectures but separate parameter sets, ensuring independent learning of task-specific and identity-specific patterns.

10.2.2 Feature Disentanglement

Given the inherent complexity of EEG signals and the coupled correlations between identity-relevant and mental task-related information, it becomes essential to eliminate the nonlinear dependencies between these two components to enable the extraction of stable cross-task identity features. To address this, we introduce hypothesis testing statistics as a measure of independence between any pair of features. The objective is to minimize the dependency between pairs of concatenated features, denoted as $f = [f_{ID}, f_T] \in \mathbb{R}^{n \times 2m}$, where n represents the number of samples, and m denotes the dimensionality of f_{ID} and f_T. Suppose A and B are random variables corresponding to any pair of feature components $f_{:,i}$ and $f_{:,j}$, where $i, j \in [1, 2m]$. Let $\mathcal{H}_A$ and $\mathcal{H}_B$ denote the associated Reproducing Kernel Hilbert Spaces (RKHS), and k_A and k_B represent their respective positive definite kernels. The cross-covariance operator $\sum_{AB}$ [14], mapping from $\mathcal{H}_A$ to $\mathcal{H}_B$ is defined by the following relationship:

$$\left\langle h_B, \sum\nolimits_{AB} h_A \right\rangle_{H_B} = \mathbb{E}_{AB}\left[h_A(A)h_B(B)\right] - \mathbb{E}_A\left[h_A(A)\right]\mathbb{E}_B\left[h_B(B)\right], \quad (10.1)$$

for all $h_A \in H_A$ and $h_B \in H_B$. To quantify the independence of two random variables in the RKHS, the Hilbert-Schmidt Independence Criterion (HSIC) is utilized. HSIC is formally defined as:

$$HSIC^{k_A,k_B}(A, B) = \left\| C_{A,B}^{k_A,k_B} \right\|_{HS}^2 = \|\Sigma_{AB}\|_{HS}^2, \quad (10.2)$$

where $\|\cdot\|_{HS}$ represents the Hilbert-Schmidt norm. Notably, it has been established that for random variables A and B, $HSIC^{k_A,k_B}(A, B) = 0$ if and only if $A \perp B$ [15], indicating independence. To mitigate the high computational cost associated with HSIC, a computationally efficient approximation based on the Frobenius norm is adopted [16]. Specifically, random functions are sampled from the function space of Random Fourier Features $\mathcal{H}_{RFF}$, defined as:

$$\mathcal{H}_{RFF} = \left\{ h : x \to \sqrt{2}\cos(\omega x + \phi) \right\}, \quad (10.3)$$

where $\omega \sim N(0, 1)$, and $\phi \sim Uniform(0, 2\pi)$. Using this framework, the partial cross-covariance matrix $\hat{\Sigma}_{AB}$ is computed as:

10.2 Disentangled Adversarial Generalization Network

$$\hat{\Sigma}_{AB} = \frac{1}{n-1} \sum_{i=1}^{n} \left[\left(\mathbf{u}(A_i) - \frac{1}{n} \sum_{j=1}^{n} \mathbf{u}(A_j) \right)^T \cdot \left(\mathbf{v}(B_i) - \frac{1}{n} \sum_{j=1}^{n} \mathbf{v}(B_j) \right) \right], \quad (10.4)$$

where,

$$\mathbf{u}(A) = \left(u_1(A), u_2(A), \ldots u_{n_A}(A) \right), u_j(A) \in \mathcal{H}_{RFF}, \forall j,$$
$$\mathbf{v}(B) = \left(v_1(B), v_2(B), \ldots v_{n_B}(B) \right), v_j(B) \in \mathcal{H}_{RFF}, \forall j.$$

Finally, the feature decorrelation loss $\mathcal{L}_{dec}$ is defined as the Frobenius norm of the partial cross-covariance matrix, computed across all pairs of concatenated features:

$$\mathcal{L}_{dec} = \sum_{1 \le i \le j \le 2m} \left\| \hat{\Sigma}_{Z_{:,i} Z_{:,j}} \right\|_F^2. \quad (10.5)$$

This loss function effectively reduces the nonlinear dependency between identity-relevant and task-relevant components, facilitating the extraction of robust and task-independent identity features.

10.2.3 Attention Mechanism with Adversarial Self-Challenging

During the training of neural networks, dominant features strongly associated with the labels are typically activated. However, in unseen test scenarios, these dominant features may have limited significance, leading to a decline in performance. To address this issue, we draw inspiration from [9] and propose an attention mechanism integrated with an adversarial self-challenging strategy. This mechanism serves as a dimensional descriptor that not only activates dominant features but also enhances the contribution of less prominent features, ensuring more robust generalization. Specifically, the disentangled identity-relevant features f_{ID} are processed through an attention mechanism $H(\cdot)$, which consists of three hidden layers. Given that f_{ID} features tend to behave like discrete random variables due to the disentanglement process, the Gumbel-Max trick [17] is employed to sample features with dimensional probabilities $\pi = H(f_{ID})$. This sampling process is defined as:

$$\arg\max_{i} \left[\log \pi_i - \log(-\log \varepsilon_i) \right]_{i=1}^{N}, \varepsilon_i \sim U(0, 1), \quad (10.6)$$

where ε_i represents a random variable sampled from a uniform distribution, and N denotes the dimensionality of the feature space. To make this sampling process continuous and differentiable, the Gumbel-Softmax trick [18] is utilized as an approximation of the $argmax$ operation. The resulting attention weights $\alpha_i \in \mathbb{R}^m$ are calculated as follows:

$$\alpha_i = \max_{i=1...N} \frac{exp\left((\log\pi_i - \log(-\log\varepsilon_i))/\tau\right)}{\sum_{j=1}^{N} exp\left((\log\pi_j - \log(-\log\varepsilon_j))/\tau\right)}, \quad (10.7)$$

where τ is the temperature parameter for the Gumbel-Softmax distribution, controlling the balance between approximation smoothness and accuracy. By leveraging the Gumbel-Softmax attention mechanism, the proposed model effectively combines the contributions of both dominant and subordinate features, allowing the network to capture more comprehensive and stable identity-relevant features. This adversarial self-challenging strategy strengthens the model's ability to generalize to unseen test data, enhancing its robustness in cross-task and cross-session brain fingerprint identification.

Subsequently, the disentangled identity-relevant features f_{ID} are further separated into the dominant identity features f_D and inferior identity features f_I based on the attention weights α. These features are defined as:

$$\begin{aligned} f_D &= a \odot f_{ID}, \\ f_I &= (1-a) \odot f_{ID}. \end{aligned} \quad (10.8)$$

where $\odot$ denotes the element-wise product operation. The dominant features f_D are fed into the identity classifier C_{ID}, while the inferior features f_I are processed by an auxiliary identity classifier C'_{ID}. Simultaneously, the disentangled task-relevant features f_T are input into the mental task classifier C_T. To train the backbone feature generator and associated classifiers, the framework employs a combination of supervision signals, including feature decorrelation loss $\mathcal{L}_{dec}$ and cross-entropy-based classification losses. These losses are formulated as follows:

$$\begin{aligned} \mathcal{L}_{cls} &= \sum_{j}^{N} E_{x \sim X_{sj}} J((G_{ID}\left(x_i^{sj}\right) \circ a), y_i^{sj}), \\ \mathcal{L}_{acls} &= \sum_{j}^{N} E_{x \sim X_{sj}} J((G_{ID}\left(x_i^{sj}\right) \circ (1-a)), y_i^{sj}), \\ \mathcal{L}_{tcls} &= \sum_{j}^{N} E_{x \sim X_{sj}} J(G_T\left(x_i^{sj}\right), c_i^{sj}), \end{aligned} \quad (10.9)$$

where $J(\cdot, \cdot)$ denotes the cross-entropy loss, G_{ID} and G_T are the identity and task feature generators, and y_i^{sj} and c_i^{sj} are the identity and task labels, respectively. The

10.3 Brain Fingerprint Identification with DAGN

training objective for DAGN is to minimize the combined loss function, formulated as an optimization problem:

$$\min_{G_{ID}, G_T, C_{ID}, C'_{ID}, C_T} \mathcal{L}_{cls} + \mathcal{L}_{acls} + \mathcal{L}_{tcls} + \lambda \mathcal{L}_{dec}, \tag{10.10}$$

where λ is a trade-off parameter that balances the influence of the feature decorrelation loss relative to the classification losses.

This optimization strategy ensures that the backbone feature generators and classifiers not only accurately predict the labels but also reduce the dependence between identity-relevant and task-relevant features. The use of attention mechanisms and separate classifiers for dominant and inferior features enhances the network's ability to generalize across tasks and sessions, thereby improving the robustness of brain fingerprint identification.

In the second step of the training process, the class identifiers are fixed, and the adversarial attention mechanism is optimized by minimizing the negative entropy of the predicted class distribution from the auxiliary identity classifier. The optimization objective is expressed as:

$$\min_{H} \mathcal{L}_{cls} - \mathcal{L}_{acls}, \tag{10.11}$$

where $\mathcal{L}_{cls}$ represents the loss for the dominant identity features, and $\mathcal{L}_{acls}$ corresponds to the loss for the inferior identity features. This adversarial training process encourages the attention mechanism to focus on extracting identity-relevant features that are both dominant and invariant across different mental tasks. It is important to note that during the test phase, only the primary identity classifier C_{ID} is used for brain fingerprint identification, ensuring an efficient and straightforward inference process. The adversarial training step enhances the robustness of the learned identity features, making them more stable and reliable for unseen tasks and sessions. The complete training procedure, including both the feature extraction and adversarial optimization stages, is outlined in Algorithm 10.1.

10.3 Brain Fingerprint Identification with DAGN

This section introduces the datasets used for experimental evaluation, followed by a detailed description of the signal preprocessing steps, data configuration, and model setup.

Algorithm 10.1 Disentangled Adversarial Generalization Network (DAGN)

Input: source labeled datasets $\mathcal{D}$, batch size n, number of training iterations T;
Output: well-trained feature generator G_{ID}, G_T, attention mechanism H, identify classifier C_{ID}, auxiliary identify classifier C'_{ID}, and task classifier C_T;
1: **for** each $t \in [1, T]$ **do**
2:   Sample a mini-batch of training samples $\{(x_i, y_i, c_i)\}_{i=1}^{n}$ from the dataset $\mathcal{D}$;
3:   Feed mini-batch training samples to the feature generators G_{ID} and G_T to obtain the identity-relevant representations $G_{ID}(x_i)$ and task-relevant representations $G_T(x_i)$;
4:   Compute the feature decorrelation loss $\mathcal{L}_{dec}$ as defined in Eq. (10.5);
5:   Input the identity-relevant representations $G_{ID}(x_i)$ into attention mechanism H to learn the attention weight a;
6:   Derive the dominant identity features f_D and inferior identity features f_I using Eq. (10.8);
7:   Feed f_D, f_I and $G_T(x_i)$ into the identify classifier C_{ID}, auxiliary identify classifier C'_{ID}, and task classifier C_T, respectively;
8:   Calculate the classification losses as defined Eq. (10.9);
9:   Update the feature generators G_{ID}, G_T, and the classifiers C_{ID}, C'_{ID}, C_T by minimizing the total loss function as defined in Eq. (10.10);
10:   Optimize the attention mechanism H by minimizing the adversarial loss defined in Eq. (10.11);
11: **end for**

10.3.1 Data Description and Pre-processing

Experiments were conducted on two multi-task EEG datasets: the Multi-Task EEG Dataset (MTED) and the 128-channel Multi-Task EEG Dataset (128-MTED), with the number of subjects for each task presented in Table 10.2. To assess the effectiveness of the proposed method, PSD features in the 3–30 Hz frequency range are extracted for each channel. This is done using a window size of 0.36 s with no overlap. The data is then segmented into 5 s samples for MTED and 15 s samples for 128-MTED. Detailed statistics on the training and testing data for each task in both datasets can be found in Table 10.2. Given the differences in the number of channels recorded in each dataset, only the common channels from the standard 10–20 system are selected for this analysis. Specifically, nine channels covering the frontal, central, parietal, and occipital lobes, Fz, F7, F8, C3, C4, P7, P8, O1, and O2, are utilized for the experiments. These channels have been demonstrated as essential for EEG-based identification in prior studies [4]. To investigate the impact of various brainprint features, we experiment with different sample lengths, frequency band ranges, and numbers of channels.

The evaluation employs two types of cross-validation: the leave-one-task-out validation and leave-test-session-out validation. In the leave-one-task-out validation, one task is excluded from the dataset and treated as the test set, while 80% of the remaining data is used for training and 20% for validation. This approach aims to assess the cross-task stability of the model. In the leave-test-session-out validation, the first 60% of sessions are used for training, while the remaining 40% is split into 80% for testing and 20% for validation. This strategy tests the model's robustness across different sessions. It is important to note that during training, no data from

10.3 Brain Fingerprint Identification with DAGN

Table 10.2 Description of dataset

Dataset	Task	Subject	Train	Test
MTED [19]	Resting state	15	2400	1065
	Watching movie clips	15	1815	1785
	Motor imaginary	15	2775	585
	Hand-grip movements	15	2775	585
128-MTED [4]	Odd ball classic	13	5217	1233
	Odd ball stereo	12	5825	474
	Imagining binary answers	7	5562	809
	Semantically opposite words	4	5900	387
	Familiar and unfamiliar words	6	5853	443
	Propper improper	8	5844	451
	Motor and mental imaginary	6	5629	725
	Passive audio	17	5513	860
	Passive audio stereo	11	5668	672
	Odd ball visual	6	6101	134
	SSVEP	12	5580	781
	Passive audio-visual	12	5618	733
	Cross-session	30	4812	2311

the test sessions is used, ensuring that the test set remains unseen and completely independent of the training phase in all experiments.

10.3.2 Baseline Approaches

This section outlines the baseline approaches used for comparison with our proposed DAGN. All these methods have been previously employed for cross-session and cross-task brain fingerprint identification.

- **EEGNet** [20]: EEGNet is a deep learning model that utilizes one-dimensional convolution to automatically extract temporal, spatial, and frequency features from EEG signals. This model has been successfully applied to brain fingerprint identification in several studies [1, 4].

- **EEGNet-PSD** [4]: While the original EEGNet model takes raw EEG data as input, our proposed DAGN uses PSD features. To provide a fair comparison, EEGNet is adapted to accept PSD features as input, following the methodology outlined in [4].
- **CNN-RNN** [1]: The CNN-RNN model combines convolutional neural networks (CNNs) with recurrent neural networks (RNNs) to extract temporal and spatial features from EEG signals. The representation S_ξ, as employed in [1], is used for comparison in our experiments.
- ix-**vector** [4]: The ix-vector approach integrates expectation-maximization (EM)-based modified-i-vectors with deep neural network (DNN)-based modified-x-vectors for the identification task. This method has been demonstrated to be effective in EEG-based identity recognition.

10.3.3 Implementation Details

All experiments were implemented using the PyTorch framework and trained in an end-to-end manner with stochastic gradient descent (SGD). The regularization parameter λ in Eq. (10.10) is dynamically adjusted throughout the training process, following the schedule $\lambda(t) = e^{-5(1-t)^2}$, where t represents the training epoch. The evaluation metrics used in this study are Rank-1 classification accuracy (ACC) and equal error rate (EER), both of which are standard measures for evaluating the performance of biometric systems.

10.3.4 Comparison of Performance with State-of-the-Art Models

This section presents a comparative performance analysis of the proposed DAGN against several baseline approaches across two datasets. The experimental results, including cross-task and cross-session validation, are summarized in Table 10.3 and Table 10.4.

In the cross-task evaluation, DAGN demonstrates substantial improvements over deep learning models that do not leverage disentangled representations. Notably, for mental tasks such as "Watching Movie Clips", "Imagining Binary Answers", "Semantically Opposite Words", and "Passive Audio-Visual", DAGN achieves significant performance gains. Specifically, the task "Passive Audio-Visual" sees an 18.14% increase in classification ACC and a 7.45% decrease in EER. When compared to subspace methods, DAGN shows the most significant improvements for tasks like "Odd Ball Visual", "SSVEP", and "Passive Audio-Visual" under the eyes-open condition. In particular, for "SSVEP", DAGN improves the ACC by 27.09% and

10.3 Brain Fingerprint Identification with DAGN

Table 10.3 Cross-task performance of DAGN and state-of-the-art models on MTED dataset [19] with ACC (%) and EER (%)

Task	EEGNet		EEGNet-PSD		CNN-RNN		DAGN (Ours)	
	ACC	EER	ACC	EER	ACC	EER	ACC	EER
Resting state	94.13 ± 2.74	0.96 ± 0.82	98.87 ± 0.53	0.43 ± 0.29	99.59 ± 0.05	0.17 ± 0.05	**99.77 ± 0.14**	**0.09 ± 0.02**
Watching movie clips	78.26 ± 8.88	4.46 ± 2.60	82.16 ± 0.32	6.54 ± 0.08	96.06 ± 0.35	1.09 ± 0.01	**98.22 ± 0.11**	**0.50 ± 0.10**
Motor imaginary	93.98 ± 3.02	2.12 ± 0.47	99.61 ± 0.32	0.09 ± 0.10	**100.00 ± 0.00**	**0.00 ± 0.00**	**100.00 ± 0.00**	**0.00 ± 0.00**
Hand-grip movements	95.22 ± 1.89	1.48 ± 0.86	99.62 ± 0.54	0.04 ± 0.03	**100.00 ± 0.00**	**0.00 ± 0.00**	**100.00 ± 0.00**	**0.00 ± 0.00**
Cross-task (Avg ± Std)	90.40 ± 8.11	3.34 ± 3.14	95.07 ± 8.61	1.78 ± 3.18	98.92 ± 1.90	0.32 ± 0.52	**99.50 ± 0.86**	**0.15 ± 0.24**

Table 10.4 Cross-task and cross-session performance of DAGN and state-of-the-art models on 128-MTED [4] dataset with ACC (%) and EER (%)

Task	EEGNet		EEGNet-PSD		CNN-RNN		ix-vector		DAGN (Ours)	
	ACC	EER	ACC	EER	ACC	EER	ACC	EER	ACC	EER
Odd ball classic	96.80	0.83	91.38	2.16	89.78	2.72	94.50	4.07	**98.12**	**0.34**
Odd ball stereo	**98.83**	**0.20**	93.35	1.52	90.58	2.31	95.30	5.72	98.05	0.67
Imagining binary answers	77.68	11.30	67.71	13.78	70.01	10.39	**96.40**	**5.36**	83.69	8.27
Semantically opposite words	82.07	**3.07**	60.94	9.21	79.67	4.47	**88.40**	11.30	86.83	3.87
Familiar and unfamiliar words	96.86	0.60	95.91	0.89	95.26	1.25	98.30	2.38	**98.87**	**0.08**
Propper improper	98.44	0.42	95.83	1.08	91.65	1.81	98.40	1.95	**98.83**	**0.15**
Motor and mental imaginary	96.88	0.65	89.23	2.32	90.21	2.00	96.92	3.77	**97.52**	**0.43**
Passive audio	69.31	21.28	65.46	22.47	66.68	22.25	**80.40**	**7.90**	71.76	17.02
Passive audio stereo	97.87	0.95	95.12	1.37	97.67	0.55	94.30	3.90	**99.01**	**0.31**
Odd ball visual	**99.48**	**0.31**	95.48	1.58	87.56	2.40	80.00	11.70	97.40	1.44
SSVEP	89.18	2.81	81.77	5.06	74.18	8.27	63.90	15.80	**90.99**	**2.42**
Passive audio-visual	68.96	9.53	71.57	6.22	63.58	9.84	81.90	11.50	**89.71**	**2.08**
Cross-task (Avg ± Std)	89.36 ± 11.77	4.33 ± 6.51	83.65 ± 13.50	5.64 ± 6.61	83.07 ± 11.75	5.69 ± 6.23	89.06 ± 10.56	7.11 ± 4.46	**92.57 ± 8.45**	**3.09 ± 4.97**
Cross-session	81.67	5.10	77.50	5.79	77.80	6.24	86.40	5.02	**90.16**	**3.17**

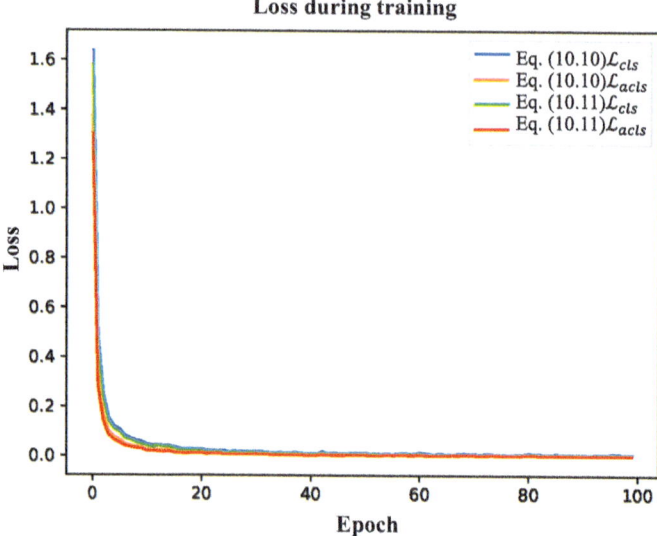

Fig. 10.2 Loss chart of cross-session validation on 128-MTED

reduces the EER by 13.38%. An insightful observation is that DAGN achieves superior performance in most leave-one-task-out cross-validation experiments, reaching the highest mean accuracies of 99.5 and 92.57%, along with the lowest EERs of 0.15 and 3.09% for the MTED and 128-MTED datasets, respectively. Furthermore, DAGN exhibits lower variability in performance across tasks when compared to baseline models, highlighting its ability to extract task-independent identity features with greater generalization capabilities.

Since the MTED dataset contains only a single session, cross-session validation was only conducted on the 128-MTED dataset, which includes multiple sessions. In this evaluation, the DAGN model significantly outperforms all comparison methods, achieving an accuracy of 90.16%. This result is 3.76% higher than the best-performing subspace technique and also shows a 1.85% reduction in EER. These results suggest that DAGN can effectively address the issue of distribution discrepancies in cross-session data and can reliably extract stable brainprint features across sessions. The loss curves for cross-session validation during training on the 128-MTED dataset are presented in Fig. 10.2. As observed, the losses $\mathcal{L}_{acls}$ and $\mathcal{L}_{cls}$ converge toward 0 during the iterative optimization process. This convergence indicates that the contributions of both dominant and inferior features to identity discrimination are well-balanced, further reinforcing the efficacy of DAGN in extracting robust identity features.

10.3.5 Ablation Experiments for the Disentangled Adversarial Generalization Network

To assess the contribution of each component in our proposed DAGN, we perform a series of ablation studies. The following variants of DAGN are evaluated to determine the impact of different components on the model's performance:

- **DAGN w/o FD & AAM**: This variant excludes both feature disentanglement (FD) and the adversarial attention mechanism (AAM);
- **DAGN w/o AAM**: This configuration omits the adversarial attention mechanism, isolating the effects of feature disentanglement;
- **DAGN w/o FD**: In this case, feature disentanglement is removed to assess the contribution of the adversarial attention mechanism in the absence of disentangled identity features.

The results of these experiments, presented in terms of classification ACC and EER on the MTED and 128-MTED datasets, are summarized in Tables 10.5 and 10.6, respectively. Across both datasets, the models "DAGN w/o FD" and "DAGN w/o AAM" outperform "DAGN w/o FD & AAM" on most tasks, highlighting the importance of both components in achieving optimal performance. These findings underscore the significance of decoupling identity information from task-specific features and learning stable, task-independent identity representations across sessions. Comparing the full DAGN model with its ablated variants reveals that DAGN consistently achieves the best results across most tasks, with notable performance gains on tasks such as "Semantically Opposite Words", "Passive Audio-Visual", and in cross-session evaluations. In particular, DAGN shows more than a 1% improvement in accuracy over the ablated models on these tasks. This reinforces the notion that DAGN's feature disentanglement and adversarial attention mechanism are key to ensuring the task-independent and cross-session stability of discriminative brainprint features.

Table 10.5 Ablation performance of DAGN on MTED [19] with ACC (%) and EER (%)

Task	DAGN w/o FD&AAM		DAGN w/o AAM		DAGN w/o FD	
	ACC	EER	ACC	EER	ACC	EER
Resting state	98.93 ± 0.11	0.36 ± 0.03	**99.57 ± 0.31**	**0.11 ± 0.03**	99.45 ± 0.41	0.13 ± 0.12
Watching movie clips	96.28 ± 0.28	0.66 ± 0.10	96.73 ± 0.25	0.70 ± 0.14	**97.49 ± 0.07**	**0.59 ± 0.08**
Motor imaginary	99.74 ± 0.15	0.02 ± 0.01	100.00 ± 0.00	0.00 ± 0.00	100.00 ± 0.00	0.00 ± 0.00
Hand-grip movements	**100.00 ± 0.00**	**0.00 ± 0.00**	100.00 ± 0.00	0.00 ± 0.00	100.00 ± 0.00	0.00 ± 0.00
Average	98.74 ± 0.10	0.26 ± 0.03	99.07 ± 0.14	0.20 ± 0.04	**99.24 ± 0.11**	**0.18 ± 0.01**

Table 10.6 Ablation performance of DAGN on 128-MTED [4] with ACC (%) and EER (%)

Task	DAGN w/o FD&AAM		DAGN w/o AAM		DAGN w/o FD	
	ACC	EER	ACC	EER	ACC	EER
Odd ball classic	97.39 ± 0.33	0.61 ± 0.11	**98.13 ± 0.55**	0.53 ± 0.13	98.10 ± 0.31	**0.46 ± 0.07**
Odd ball stereo	**97.46 ± 0.27**	0.94 ± 0.23	97.27 ± 0.28	**0.86 ± 0.08**	97.20 ± 0.33	1.14 ± 0.20
Imagining binary answers	79.60 ± 1.65	9.17 ± 0.63	81.20 ± 1.68	**8.57 ± 1.90**	**82.09 ± 0.41**	8.72 ± 1.13
Semantically opposite words	83.78 ± 1.24	4.08 ± 0.39	83.53 ± 4.62	**3.81 ± 0.67**	**84.37 ± 1.19**	4.63 ± 0.21
Familiar and unfamiliar words	98.89 ± 0.32	0.23 ± 0.13	98.96 ± 0.21	0.19 ± 0.02	**99.18 ± 0.11**	**0.13 ± 0.06**
Propper improper	97.85 ± 0.16	0.52 ± 0.08	**98.31 ± 0.18**	**0.45 ± 0.10**	98.24 ± 0.16	0.49 ± 0.13
Motor and mental imaginary	96.92 ± 0.82	**0.52 ± 0.05**	96.65 ± 0.45	0.81 ± 0.20	**97.35 ± 0.54**	0.61 ± 0.23
Passive audio	69.75 ± 0.50	21.28 ± 0.80	**70.43 ± 0.42**	21.74 ± 0.67	70.05 ± 0.69	**18.44 ± 0.49**
Passive audio stereo	98.82 ± ± ± 0.47	0.28 ± 0.04	**99.15 ± 0.11**	0.29 ± 0.04	98.91 ± 0.29	**0.27 ± 0.17**
Odd ball visual	96.88 ± 0.74	0.50 ± 0.32	97.05 ± 1.49	**0.32 ± 0.18**	97.05 ± 1.07	0.58 ± 0.04
SSVEP	87.02 ± 0.61	3.00 ± 0.36	**88.86 ± 0.64**	2.57 ± 0.13	87.90 ± 0.99	**2.42 ± 1.51**
Passive audio-visual	85.02 ± 1.11	3.15 ± 0.24	86.93 ± 0.15	3.06 ± 0.27	**88.15 ± 0.28**	**2.51 ± 0.34**
Average	90.78 ± 0.34	3.69 ± 0.19	91.37 ± 0.40	3.60 ± 0.20	**91.55 ± 0.17**	**3.37 ± 0.34**
Cross-session	87.86 ± 0.31	3.54 ± 0.10	88.30 ± 0.19	3.67 ± 0.15	**88.86 ± 0.54**	**3.23 ± 0.10**

Feature disentanglement removes spurious correlations between identity and mental task information, leading to more robust task-independent brainprint features. Additionally, it mitigates distributional differences in multi-task features across sessions, enhancing the model's generalization capabilities through the self-challenging adversarial mechanism.

To further validate the effectiveness of FD and the AAM, we compute the cosine similarity between the identity features, f_{ID}, extracted from the training and test sets across different mental tasks. The results, presented in Table 10.7, demonstrate that the identity features captured by DAGN on the training and test sets exhibit a high degree of similarity, particularly on the 128-MTED dataset, which contains multi-task and multi-session EEG data. On the other hand, "DAGN w/o AAM" performs better on the MTED dataset, which consists of data from a single session. The

10.3 Brain Fingerprint Identification with DAGN

Table 10.7 Comparison of cosine similarity between the identity features f_{ID} from the training and test sets

Dataset	Task	DAGN w/o AAM	DAGN w/o FD	DAGN (Our)
MTED [19]	Resting state	**0.9958**	0.9947	0.9948
	Watching movie clips	**0.9938**	0.9920	0.9928
	Motor imaginary	**0.9966**	0.9964	0.9960
	Hand-grip movements	**0.9964**	0.9956	0.9962
128-MTED [4]	Odd ball classic	0.8855	0.9371	**0.9616**
	Odd ball stereo	0.8731	0.9337	**0.9579**
	Imagining binary answers	0.8964	0.9042	**0.9048**
	Semantically opposite words	**0.9348**	0.9188	0.9310
	Familiar and unfamiliar words	0.9663	0.9784	**0.9872**
	Propper improper	0.9016	0.9482	**0.9711**
	Motor and mental imaginary	0.9802	0.9790	**0.9837**
	Passive audio	0.8973	0.9440	**0.9688**
	Passive audio stereo	0.8933	0.9533	**0.9780**
	Odd ball visual	0.8774	0.9202	**0.9543**
	SSVEP	0.8086	0.8878	**0.9148**
	Passive audio-visual	0.8330	0.8807	**0.9275**

adversarial attention mechanism is specifically designed to balance the contribution of identity features across different dimensions in the presence of cross-session data with varying distributions. It enhances model robustness when dealing with unseen sessions. In contrast, the effect of AAM is less pronounced in the MTED dataset, where the data distribution differences between training and test sets are minimal due to the use of a single session. These results suggest that while feature disentanglement successfully isolates identity-related features from task-related information, the AAM plays a critical role in enhancing the model's generalization ability, especially when handling multi-session EEG data with diverse distributions. By combining FD and AAM, DAGN effectively extracts robust identity features that generalize well across different tasks and sessions.

To evaluate the optimization of the adversarial attention mechanism, we compare the Euclidean distances between the dominant identity features f_D and inferior identity features f_I, as extracted by DAGN and the ablated "DAGN w/o AAM" model. As shown in Table 10.8, the distance between f_I and f_D is greater for DAGN than

Table 10.8 Comparison of euclidean distance between the dominant identity features f_D and inferior identity features f_I

Dataset	Task	DAGN w/o AAM	DAGN (Our)
MTED [19]	Resting state	6.8063	**7.3699**
	Watching movie clips	6.9105	**7.4168**
	Motor imaginary	7.2863	**7.3805**
	Hand-grip movements	6.7615	**7.3459**
128-MTED [4]	Odd ball classic	4.9295	**7.1538**
	Odd ball stereo	4.0775	**6.8832**
	Imagining binary answers	4.5108	**7.0963**
	Semantically opposite words	3.6094	**6.3330**
	Familiar and unfamiliar words	4.2014	**7.2801**
	Propper improper	4.1584	**6.6159**
	Motor and mental imaginary	4.0402	**6.7878**
	Passive audio	3.8185	**5.8153**
	Passive audio stereo	4.1118	**7.0910**
	Odd ball visual	4.0500	**6.1312**
	SSVEP	3.9006	**6.5839**
	Passive audio-visual	3.9945	**6.2033**

for "DAGN w/o AAM", indicating that the AAM optimized with Eq. (10.11), successfully enhances the distinction between dominant and inferior identity features. This demonstrates the efficacy of the AAM in highlighting the inferior features while preserving the integrity of dominant identity features for accurate recognition.

Finally, to visually investigate the effectiveness of feature disentanglement, we use t-SNE [21] to project the extracted brainprint features into two dimensions. The embeddings of the brainprint features captured by DAGN are shown in Fig. 10.3a–e. Remarkably, the features extracted by DAGN exhibit better separation between classes compared to those from the contrastive model. Additionally, the distributions of features from the training data are notably more aligned with those from the test data, indicating that DAGN effectively captures task-independent identity features that generalize well to unseen data.

10.3.6 Effect of Sample Length

The length of the sample directly influences the amount of identity-related information available for feature extraction, with longer sample durations typically providing

10.3 Brain Fingerprint Identification with DAGN

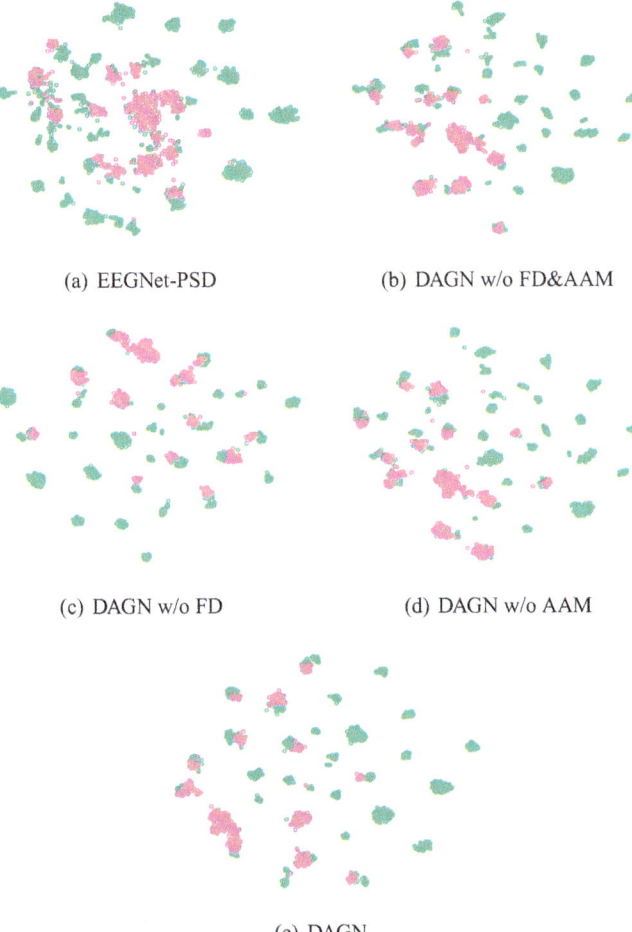

Fig. 10.3 The t-SNE visualization of latent representations for test task Passive Audio-Visual with 12 subjects captured by **a** EEGNet-PSD, **b** DAGN w/o FD&AAM, **c** DAGN w/o FD, **d** DAGN w/o AAM, **e** DAGN. The green dots indicate the representations of the training set, and the pink dots indicate the representations of the test set

more comprehensive identity features. However, this comes at the cost of increased computational complexity, which could pose practical limitations for real-time applications. In this section, we evaluate the impact of varying sample lengths on the performance of brain fingerprint identification. Specifically, EEG signals are segmented into sample lengths of 5, 10, 15, and 30 s, and the corresponding experimental results for different test tasks are presented in Fig. 10.4.

From the results, we observe that shorter time slices, such as 5 and 10 s, generally lead to superior performance in the single-session dataset. Notably, the performance

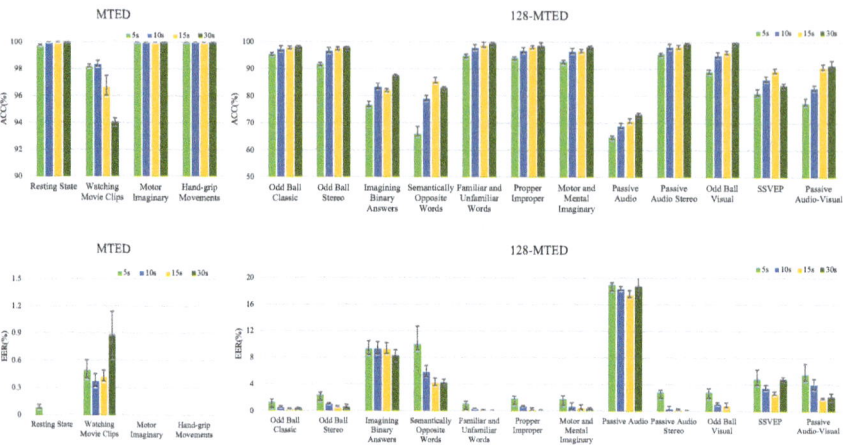

Fig. 10.4 Performance of proposed DAGN on samples with varying length

on the "watching movie clips" task slightly deteriorates with the shorter sample durations. This performance drop can likely be attributed to the limited number of training samples available, which constrains the model's ability to fully capture the identity-related features from these shorter time windows. Conversely, when evaluating the model on the cross-session dataset, we observe that 5 s samples consistently perform the worst across all tasks. As the sample length increases, the model's performance improves steadily. However, this improvement is not linear, and the growth rate of performance starts to taper off as the sample length exceeds 10 s. These results suggest that while longer sample lengths help the model capture more robust and stable identity features, the incremental gain in performance diminishes beyond a certain point.

In summary, these findings indicate that our model is capable of effectively extracting stable identity information from relatively short-time slices, and that a sample length of 10 s strikes a good balance between performance and computational efficiency. This suggests that, for real-world applications, EEG segments longer than 10 s can be used to extract stable, task-independent brainprint features, while shorter time slices (5 or 10 s) can be sufficient to achieve high accuracy, particularly when training data is limited.

10.3.7 Effect of the Different Frequency Bands

EEG signals are generated by spontaneous, rhythmic neural oscillations that span several frequency bands, including theta (3–8 Hz), alpha (8–12 Hz), beta (12–30 Hz), and gamma (30–45 Hz). These bands are commonly associated with distinct neural activities and cognitive states, making it crucial to investigate their contributions to

10.3 Brain Fingerprint Identification with DAGN

the stability and robustness of brainprint features across different tasks and sessions. To this end, extensive experiments are conducted to evaluate the performance of the proposed model across these various frequency bands. The experimental results obtained from both datasets are summarized in Table 10.9.

The comparison of performance across the four frequency bands reveals that the 3–8 Hz band, corresponding to the theta band, plays a crucial role in maintaining the stability of brainprint features across both tasks and sessions. While the 30–45 Hz gamma band yielded superior performance in certain tasks within the 128–MTED dataset, it did not consistently outperform the 3–8 Hz band. In fact, the theta band

Table 10.9 Performance of DAGN on different frequency bands with ACC (%) and EER (%)

Dataset	Task	3–8 Hz		8–12 Hz		12–30 Hz		30–45 Hz		3–45 Hz	
		ACC	EER	ACC	EER	ACC	EER	ACC	EER	ACC	EER
MTED [19]	Resting state	99.80[a]	0.08	99.80[a]	0.11[a]	99.88[a]	0.10	99.80[a]	0.16[a]	**100.00**	**0.00**
	Watching Movie Clips	**98.35**	**0.20**	97.65	0.47	97.92	0.59	97.78	0.40	97.48	0.42
	Motor imaginary	100.00	0.00	100.00	0.00	100.00	0.00	100.00	0.00	100.00	0.00
	Hand-grip movements	100.00	0.00	100.00	0.00	100.00	0.00	100.00	0.00	100.00	0.00
	Average	99.54	0.07	99.36	0.15	99.45	0.17	99.40	0.14	99.37	0.11
128-MTED [4]	Odd ball classic	98.02[a]	0.44	97.40[a]	0.55	97.29[a]	0.73[a]	97.92	0.52	**99.06**	**0.38**
	Odd ball stereo	97.33[a]	0.91	97.07[a]	1.27	96.88	1.01	96.29	1.11	96.35	2.50
	Imagining binary answers	**82.31**	9.62	81.44	9.46	81.38	**9.37**	78.20	10.33	81.83	9.50
	Semantically opposite words	**84.67**	5.18	82.14	6.28	82.96	**5.48**	71.41[a]	8.76	80.43	5.59
	Familiar and unfamiliar words	99.47	0.04	99.63	**0.03**	98.95	0.17	99.33	0.06	**99.70**	0.06
	Propper improper	98.18[a]	0.49	97.85[a]	0.42	97.79[a]	0.52[a]	98.57	0.27	**98.90**	**0.23**
	Motor and mental imaginary	94.96[a]	0.86	96.44[a]	0.76[a]	96.05[a]	1.05[a]	98.22	0.36	**98.44**	**0.29**
	Passive audio	70.27	18.03	69.20	18.03	69.86	18.51	67.60	20.19[a]	**70.50**	**17.82**
	Passive audio stereo	98.63[a]	0.31	98.25[a]	0.27	98.96	0.23	98.48[a]	0.39	**99.20**	**0.16**
	Odd Ball visual	96.53	0.84	96.01	0.77	96.01	1.11	95.66	1.22	**97.23**	**1.16**
	SSVEP	86.38	3.35	88.02	3.27	87.98	3.26	**89.82**	2.43	89.34	2.77
	Passive audio-visual	89.34	2.49	86.62[a]	3.38[a]	87.84[a]	2.82[a]	89.71[a]	2.37[a]	**93.97**	**1.16**
	Average	91.34	3.55	90.84[a]	3.71[a]	91.00[a]	3.69	90.10[a]	4.00[a]	**92.08**	**3.47**

[a] Is defined as follows $p < 0.05$

demonstrated strong identity discriminability across both datasets, even in cases where the gamma band showed improved performance for specific tasks. This suggests that lower-frequency bands, such as the theta band, may be more robust for identity recognition, contributing to more stable brainprint features. A direct comparison between the performance of the full-band EEG data and the 3–8 Hz band reveals that, on the MTED dataset, the average classification ACC improved by 99.54%, and the EER decreased by 0.07%, although these differences were not statistically significant. Notably, across both datasets, the ACC for all frequency bands exceeded 90%, further reinforcing the notion that brainprint features are distributed across different frequency ranges. However, the results suggest that low-frequency bands, particularly the 3–8 Hz band, tend to be more associated with identity-related features that remain stable across sessions, while higher frequency bands such as alpha, beta, and gamma are more strongly linked to task-related cognitive processes.

In conclusion, these findings indicate that the stability of brainprint features is highly influenced by the frequency band used, with the theta band (3–8 Hz) providing a more stable and task-independent representation of identity across sessions. While higher frequency bands may offer additional task-specific information, the theta band's contribution to stable identity features makes it particularly valuable for cross-session and cross-task brain fingerprint identification.

10.3.8 Effect of the Number of Electrodes

The electrode configuration in EEG data provides valuable spatial information about the brain, making it essential to investigate how different brain regions contribute to the effectiveness of brain fingerprint identification. This section explores the impact of varying electrode placements on performance across different tasks, with particular emphasis on understanding how the number and location of electrodes influence the extraction of task-independent brainprint features. In our experiments, the EER for the "Motor Imagery" and "Hand-grip Movements" tasks was 0% across different electrode sets, meaning no significant differences were observed for these tasks. Consequently, the results for these tasks are not included in Fig. 10.5. Based on the dataset details provided in Table 10.2, it is likely that the training data for these tasks is about five times larger than the test data. As a result, only a limited number of electrodes are necessary to extract stable, task-independent brainprint features from the larger training set in a single-session scenario.

In the experiments, seven different electrode configurations were evaluated, which include electrodes from the Frontal, Central, Parietal, Temporal, and Occipital lobes. The electrode sets are illustrated in Fig. 10.5, and each set includes electrodes from various brain regions. The experimental results, shown in Fig. 10.5, reveal several insightful trends. Notably, the 5-channel-2 configuration outperformed the 5-channel-1 set on most tasks, suggesting that electrodes placed over the Parietal and Occipital lobes are more critical for brain fingerprint identification. Furthermore, there was no significant performance difference between the 8-channel-1 and

10.3 Brain Fingerprint Identification with DAGN

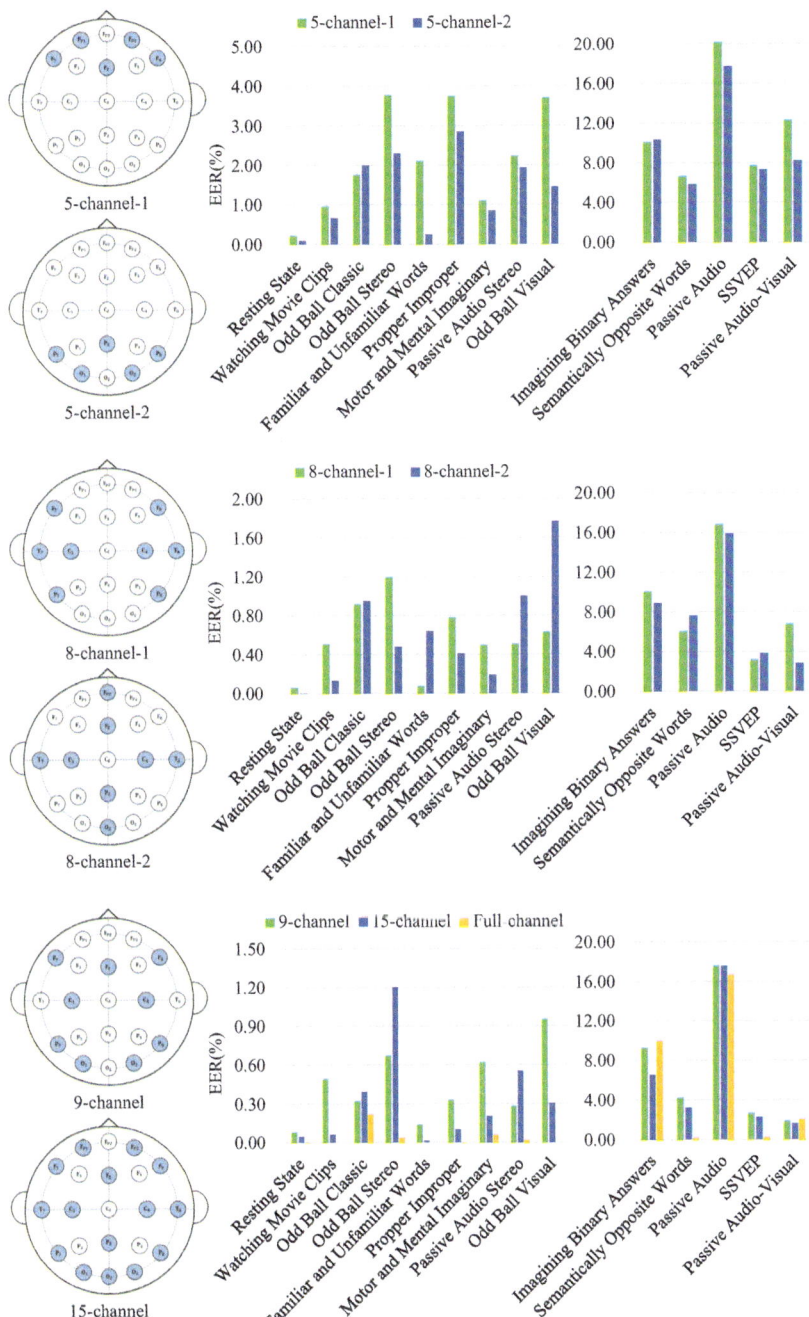

Fig. 10.5 Performance of DAGN with different number of electrodes. The channel map on the left highlights the sensors selected for each set

8-channel-2 configurations, which suggests that adding electrodes in the Temporal lobe may not contribute significantly to task-independent brainprint extraction. This observation highlights the importance of the Parietal and Occipital regions in enhancing the stability and accuracy of brain fingerprint features. As the number of electrodes increased, model performance consistently improved. The best results across most tasks were achieved with the full-channel configuration, which includes 62 and 128 electrodes, providing comprehensive coverage of the Frontal, Central, Parietal, Temporal, and Occipital lobes. These findings suggest that incorporating additional electrodes from multiple brain regions contributes to a more robust and accurate extraction of brainprint features, particularly for cross-task and cross-session identification.

In conclusion, our results underscore the importance of electrode placement and the number of electrodes in optimizing brain fingerprint identification. The Parietal and Occipital lobes play a crucial role in ensuring the stability and discriminability of identity features, while the gradual improvement in performance with increased electrode coverage highlights the advantages of leveraging a more comprehensive spatial representation of brain activity.

10.4 Conclusion

In this chapter, we introduced a novel method for brain fingerprint identification based on disentangled representations learning and adversarial generalization, aimed at extracting stable, task-independent brainprint features across various tasks and sessions. Our approach tackles the challenge of spurious correlations between task-relevant and identity-relevant information by explicitly decorrelating these factors. This enables the identification of task-independent brainprint features, which are crucial for reliable and generalizable brain fingerprint recognition. Additionally, we incorporated an adversarial self-challenging attention mechanism that enhances the model's ability to leverage discriminative identity features, thereby improving the robustness of brainprint features across different sessions.

The proposed method demonstrates significant advantages over state-of-the-art models, especially in terms of generalization across tasks and sessions. Our approach's ability to handle varying sample lengths, frequency bands, and electrode configurations contributes to its robustness and scalability in practical applications of brain fingerprint identification. These findings confirm that our model can effectively separate identity-related information from task-related noise, ensuring consistent performance even in the presence of session and task variability.

However, despite its strengths, there are inherent limitations in the current framework. The multi-task, multi-session datasets employed in this study, though comprehensive, still feature a relatively small number of subjects. This restricts the method's ability to fully assess its scalability and generalization across larger, more diverse populations. Moreover, while the method performs well across several configurations, there is still potential for improvement in extreme conditions such as highly noisy or cross-domain scenarios.

Looking forward, the proposed method will be extended to larger, more challenging datasets, which include more subjects and diverse tasks across multiple sessions. Such expansion will allow for further refinement and validation of its effectiveness, particularly in addressing the complexities posed by real-world applications in biometric identification. The results from these future studies will be crucial for improving the adaptability of the model in practical environments, such as high-security identity authentication systems, healthcare, and human-computer interactions.

References

1. Maiorana E (2020) Deep learning for EEG-based biometric recognition. Neurocomputing 410:374–386
2. Wang M, El-Fiqi H, Hu J, Abbass HA (2019) Convolutional neural networks using dynamic functional connectivity for eeg-based person identification in diverse human states. IEEE Trans Inf Forensics Secur 14(12):3259–3272
3. Jin X, Tang J, Kong X, Peng Y, Cao J, Zhao Q, Kong W (2020) CTNN: a convolutional tensor-train neural network for multi-task brainprint recognition. IEEE Trans Neural Syst Rehabil Eng 29:103–112
4. Kumar MG, Narayanan S, Sur M, Murthy HA (2021) Evidence of task-independent person-specific signatures in EEG using subspace techniques. IEEE Trans Inf Forensics Secur 16:2856–2871
5. Nir Y, Mukamel R, Dinstein I, Privman E, Harel M, Fisch L, Gelbard-Sagiv H, Kipervasser S, Andelman F, Neufeld MY et al (2008) Interhemispheric correlations of slow spontaneous neuronal fluctuations revealed in human sensory cortex. Nat Neurosci 11(9):1100–1108
6. Addante RJ, Watrous AJ, Yonelinas AP, Ekstrom AD, Ranganath C (2011) Prestimulus theta activity predicts correct source memory retrieval. Proc Natl Acad Sci 108(26):10702–10707
7. O'Connell RG, Dockree PM, Bellgrove MA, Turin A, Ward S, Foxe JJ, Robertson IH (2009) Two types of action error: electrophysiological evidence for separable inhibitory and sustained attention neural mechanisms producing error on go/no-go tasks. J Cogn Neurosci 21(1):93–104
8. Wang H, He Z, Lipton ZC, Xing EP (2019) Learning robust representations by projecting superficial statistics out. In: International conference on learning representations
9. Huang Z, Wang H, Xing EP, Huang D (2020) Self-challenging improves cross-domain generalization. In: European conference on computer vision. Springer, Berlin, pp 124–140
10. Rahimi A, Recht B (2007) Random features for large-scale kernel machines. In: Platt J, Koller D, Singer Y, Roweis S (eds) Advances in neural information processing systems, vol 20. Curran Associates, Inc
11. Zheng W-L, Bao-Liang L (2015) Investigating critical frequency bands and channels for EEG-based emotion recognition with deep neural networks. IEEE Trans Auton Ment Dev 7(3):162–175
12. Peng Y, Qin F, Kong W, Ge Y, Nie F, Cichocki A (2022) GFIL: a unified framework for the importance analysis of features, frequency bands, and channels in EEG-based emotion recognition. IEEE Trans Cogn Dev Syst 14(3):935–947
13. Liu K, Yang M, Zhuliang Yu, Wang G, Wei W (2023) Fbmsnet: a filter-bank multi-scale convolutional neural network for EEG-based motor imagery decoding. IEEE Trans Biomed Eng 70(2):436–445

14. Fukumizu K, Bach FR, Jordan MI (2004) Dimensionality reduction for supervised learning with reproducing kernel Hilbert spaces. J Mach Learn Res 5(Jan):73–99
15. Bahng H, Chun S, Yun S, Choo J, Oh SJ (2020) Learning de-biased representations with biased representations. In: International conference on machine learning. PMLR, pp 528–539
16. Zhang X, Cui P, Xu R, Zhou L, He Y, Shen Z (2021) Deep stable learning for out-of-distribution generalization. In: Proceedings of the IEEE/CVF conference on computer vision and pattern recognition, pp 5372–5382
17. Maddison CJ, Tarlow D, Minka T (2014) A* sampling. Advances in neural information processing systems, vol 27
18. Jang E, Gu S, Poole B (2017) Categorical reparametrization with gumble-softmax. In: International conference on learning representations (ICLR 2017). OpenReview.net
19. Kong X, Kong W, Fan Q, Zhao Q, Cichocki A (2018) Task-independent EEG identification via low-rank matrix decomposition. In: 2018 IEEE international conference on bioinformatics and biomedicine (BIBM). IEEE, pp 412–419
20. Lawhern VJ, Solon AJ, Waytowich NR, Gordon SM, Hung CP, Lance BJ (2018) Eegnet: a compact convolutional neural network for EEG-based brain-computer interfaces. J Neural Eng 15(5):056013–056013
21. Donahue J, Jia Y, Vinyals O, Hoffman J, Zhang N, Tzeng E, Darrell T (2014) Decaf: a deep convolutional activation feature for generic visual recognition. In: Proceedings of the 31th international conference on machine learning, vol 32, pp 647–655

Open Access This chapter is licensed under the terms of the Creative Commons Attribution-NonCommercial-NoDerivatives 4.0 International License (http://creativecommons.org/licenses/by-nc-nd/4.0/), which permits any noncommercial use, sharing, distribution and reproduction in any medium or format, as long as you give appropriate credit to the original author(s) and the source, provide a link to the Creative Commons license and indicate if you modified the licensed material. You do not have permission under this license to share adapted material derived from this chapter or parts of it.

The images or other third party material in this chapter are included in the chapter's Creative Commons license, unless indicated otherwise in a credit line to the material. If material is not included in the chapter's Creative Commons license and your intended use is not permitted by statutory regulation or exceeds the permitted use, you will need to obtain permission directly from the copyright holder.

Chapter 11
Summary

In this book, we have explored the concept and various methodologies of brain fingerprint identification, with a focus on addressing key challenges in the field. While brain fingerprint identification has demonstrated significant potential for secure and continuous personal identification, several issues remain unresolved, particularly in ensuring the stability of brainprint features across sessions and in broadening the applicability of brain fingerprint identification to task-independent scenarios. The primary findings and contributions of this book are summarized below:

1. **Addressing Cross-Session Instability**
   One of the major challenges in brain fingerprint identification is the difficulty in extracting stable and reliable brainprint features across different sessions. As EEG signals are highly susceptible to noise, external environmental factors, and fluctuations in cognitive states, achieving cross-session consistency is a critical hurdle. This book has highlighted several methods and strategies to improve the robustness of brain fingerprint identification systems, such as Multi-scale Graph Neural Network and joint Disentangled Representation. Future work should focus on further refining these approaches to ensure that brainprint features remain stable and invariant across different recording sessions.

2. **Overcoming Task-Specific Limitations**
   Many existing brain fingerprint identification methods rely on specific cognitive tasks or active cooperation from the subject, limiting their practical utility in real-world applications. We have reviewed various task-specific paradigms and discussed their constraints, particularly in scenarios where external stimuli or particular mental states are required. To address this limitation, we propose the exploration of task-independent brain fingerprint identification methods, such as Brain Network, Low-Rank, Matrix Decomposition, Residual Multi-scale Neural Network, and Convolutional Tensor-Train Neural Network. Experimental results show that the proposed methods achieve state-of-the-art performance on multiple specific tasks and multi-task datasets. These would allow for the identification of

individuals based on more generalized and consistent neural patterns, regardless of the specific mental task being performed. This would significantly enhance the versatility and scalability of brain fingerprint identification systems in various settings.

3. **Improving Model Generalization**
   In the existing literature, there is often a reliance on EEG data from specific tasks or sessions, which can lead to the development of models that lack generalizability. Pseudo-correlations between identity-related features and non-identity-related features, especially under different cognitive states or environmental influences, pose a significant challenge to the robustness of identification systems. This book has examined methods to address this issue, including Attention Neural Network with Domain Adaptation Learning and Disentangled Adversarial Generalization Network, to increase the generalizability of brain fingerprint recognition systems.

This book highlights the immense practical potential of brain fingerprint identification in a wide range of fields, including high-security environments, healthcare, finance, and the metaverse. As the technology matures, we can expect to see widespread adoption of brain fingerprint identification systems that are secure, non-invasive, and resistant to spoofing or coercion. The continued advancement of brain fingerprint identification will likely play a transformative role in the evolution of biometric security systems, personal identification methods, and cognitive research.

Open Access This chapter is licensed under the terms of the Creative Commons Attribution-NonCommercial-NoDerivatives 4.0 International License (http://creativecommons.org/licenses/by-nc-nd/4.0/), which permits any noncommercial use, sharing, distribution and reproduction in any medium or format, as long as you give appropriate credit to the original author(s) and the source, provide a link to the Creative Commons license and indicate if you modified the licensed material. You do not have permission under this license to share adapted material derived from this chapter or parts of it.

The images or other third party material in this chapter are included in the chapter's Creative Commons license, unless indicated otherwise in a credit line to the material. If material is not included in the chapter's Creative Commons license and your intended use is not permitted by statutory regulation or exceeds the permitted use, you will need to obtain permission directly from the copyright holder.

Chapter 12
Future Directions

The work presented in this book addresses key challenges in EEG-based brain fingerprint identification. In this chapter, we introduce several foundational contributions that have laid the groundwork for future exploration in this area. This chapter focuses on the potential avenues for further research based on these preliminary studies and other emerging needs in the field.

1. **Small Sample Learning for EEG Biometrics**
   Building on these advancements, small-sample learning remains a critical frontier for brain fingerprint systems. Many practical applications of EEG biometrics face the challenge of limited data availability. Addressing this issue through techniques such as transfer learning, few-shot learning, and synthetic data generation can significantly enhance the scalability of EEG-based identification systems. Future research should prioritize improving the performance of these systems when large datasets are unavailable, ensuring robustness across diverse environments and cognitive tasks.
2. **Robustness Against Adversarial Attacks**
   With the rise of deep learning in EEG-based biometric systems, the susceptibility of these models to adversarial attacks has become a pressing concern. A particularly worrisome challenge is the potential manipulation of EEG signals, which could undermine identification accuracy. One of the pioneering studies introduced an innovative adversarial attack method that targets both time-domain and frequency-domain EEG signals using wavelet transforms. This dual-domain approach enhances the imperceptibility of the attacks while maintaining high attack efficiency. Future research should prioritize developing robust defense mechanisms to counter these vulnerabilities. Strengthening adversarial training strategies and leveraging the unique properties of EEG signals could improve the security of EEG-based biometric systems, particularly in scenarios where accuracy and resilience to attacks are paramount.

3. **Privacy-Preserving EEG Systems**
   As EEG-based biometrics become more widely adopted, privacy concerns regarding the handling of sensitive EEG data have intensified. Traditional EEG systems, particularly those relying on cloud-based computations, risk exposing personal brain data to potential misuse on untrusted servers. A significant advancement in this area introduces a privacy-preserving EEG biometric identification system using homomorphic encryption. This system enables secure processing of EEG data on untrusted servers without revealing sensitive raw data. The method substantially reduces computational and storage demands by applying encryption to brainprint features rather than the full EEG signals. Future research should focus on optimizing encryption techniques to enhance system efficiency while ensuring that privacy concerns are adequately addressed, particularly in applications where real-time performance is critical.

These early contributions set the stage for significant advancements in brain fingerprint identification. Future research can further enhance the reliability and applicability of EEG biometrics by addressing adversarial robustness, privacy concerns, and the challenges of small-sample learning. These foundational works provide a solid base for expanding the scope of brain fingerprint identification and ensuring its secure, efficient, and scalable deployment in real-world applications.

Open Access This chapter is licensed under the terms of the Creative Commons Attribution-NonCommercial-NoDerivatives 4.0 International License (http://creativecommons.org/licenses/by-nc-nd/4.0/), which permits any noncommercial use, sharing, distribution and reproduction in any medium or format, as long as you give appropriate credit to the original author(s) and the source, provide a link to the Creative Commons license and indicate if you modified the licensed material. You do not have permission under this license to share adapted material derived from this chapter or parts of it.

The images or other third party material in this chapter are included in the chapter's Creative Commons license, unless indicated otherwise in a credit line to the material. If material is not included in the chapter's Creative Commons license and your intended use is not permitted by statutory regulation or exceeds the permitted use, you will need to obtain permission directly from the copyright holder.

The manufacturer's authorised representative in the EU is Springer Nature Customer Service Centre GmbH, Europaplatz 3, 69115 Heidelberg, Germany. If you have any concerns regarding our products, please contact ProductSafety@springernature.com

Printed and bound by CPI Group (UK) Ltd, Croydon, CR0 4YY

26/03/2026

02078941-0012